51 ... ns with the Wise Old Fire Protection Engineer

Jeffrey Moore, PE, FSFPE

FIREMARK PRESS

Notice Concerning Liability

The publication of this work is to circulate educational information and opinion among those concerned about fire safety in industrial, commercial, and institutional facilities. While every effort was made to provide a work of high quality, neither the author nor publisher guarantees the accuracy or completeness of or assumes any liability in connection with the information and opinions contained herein. The publisher and author shall in no event be liable for any personal injury, property, or damage of any nature directly or indirectly resulting from the publication, use of, or reliance upon this work. The ideas and opinions expressed are solely those of the author.

This book is not intended as and should not be used as a design guide. Diagrams and illustrations are for educational and explanatory purposes only. They do not necessarily show all required equipment and details of a particular fire protection system or feature. This text does not purport to list all possible hazards in a facility, that other hazards do not exist, or that compliance with any of the recommendations or opinions expressed will make any facility safe or in compliance with any law, rule, or regulation.

References to fire and life safety code requirements were current as of the date of preparation but are for illustrative purposes only. Published codes and standards change on a regular cycle. Always consult the most recent edition of the code or standard applicable to the situation under examination. The facility owner and operator retain full responsibility for the safety of all operations. This work is published with the understanding that the author and publisher are supplying information and opinion but are not attempting to render engineering or other professional services. If such services are required, always seek the assistance of a qualified licensed design professional.

Print version 09062022

Print (Softbound) ISBN: 978-1-7379622-9-8
Library of Congress Control Number: 2022914528

To all the Wise Old Fire Protection Engineers who patiently serve as teachers and mentors to the Brash Young Fire Protection Engineers eager to learn the science and practice of the profession, and to Walter W. Maybee, PE, FSFPE, the original Wise Old Fire Protection Engineer.

TABLE OF CONTENTS

Author's Note

In my over forty-year fire protection engineering career, I've been fortunate to have many "Wise Old Fire Protection Engineers" as teachers and mentors. I learned and continue to learn much from them through discussions and casual conversations at meetings, conferences, project sites, and simply during a phone call or over dinner when time allows. These discussions provided information from the professional experience of my mentors, and I learned much of the most valuable and most practical knowledge I've acquired and applied to various fire protection problems during my career. Moreover, I've found such honest real-world conversations are still one of the most effective means of both teaching and learning.

I continue to learn similarly and try to emulate my mentors by passing the knowledge on whenever possible. However, including conversational information about unique situations and the history behind many fundamental fire protection concepts is challenging to include in a textbook since this information doesn't necessarily fit into the nice, neat black and white topics and categories found in a code book or fire protection handbook. How to include this valuable information while preparing the third edition of ***Industrial Fire Control Concepts*** posed a problem.

My solution was to borrow an idea from Walter W. Maybee, PE, FSFPE, former head of fire protection engineering for the US Department of Energy. He was a Factory Insurance Association (FIA/IRI) alumnus who, much like me, received a firm foundation in industrial and facility fire protection in the FIA/IRI's Fire Safety Laboratory early in his career. Walt was an admirer of the first edition of ***Industrial Fire Control Concepts***, writing that it was "*exactly the kind of book he tried to develop after attending the FIA [IRI] training courses early in his career.*"

For years Walt published an unofficial newsletter for federal fire protection engineers. Called "804"—the numerical designator of the job classification for federal fire protection engineers, he often provided little tidbits of trivia, or historical information, through brief three or four line conversations between a "Brash Young Fire Protection Engineer (BYFPE)" and his mentor, the "Wise Old Fire Protection Engineer (WOFPE)." I found his idea to be an effective and entertaining means of communicating key concepts.

I expanded Walt's idea for use in the third edition of ***Industrial Fire Control Concepts***. Throughout that text, I provided conversations between the BYFPE and his or her mentor, the WOFPE. These conversations offering helpful background information and explanation of topics often raising questions during the training courses and seminars I've taught were well received by reviewers and readers of ***Industrial Fire Control Concepts***. In fact, several suggested I collect and publish these conversations separately. That's why this book exists.

These conversations are intended as a means of communicating practical, down-to-earth fire protection concepts and to dispel some common fire protection myths and misconceptions. Most of the conversations between the Wise Old Fire Protection Engineer (WOFPE) and the Brash Young Fire Protection Engineer (BYFPE) in this book originally appeared in the third edition of ***Industrial Fire Control Concepts***. I added a few other conversations that I had on the shelf, but which didn't make it into the ***Industrial Fire Control Concepts*** book.

Apart from Chapter 13, the chapters in this book and the conversations in each chapter correspond to where they appeared in the twelve chapters of ***Industrial Fire Control Concepts***. I added Chapter 13 to address topics related to the professional practice of fire protection engineering, a subject not specifically addressed in the ***IFCC*** book.

And for anyone who thinks I claim to be the "Wise Old Fire Protection Engineer," I don't. The WOFPE/BYFPE interactions are simply a literary device to convey information in a concise and entertaining manner. Like probably most everyone else in the field today, somedays you feel like the WOFPE and somedays you feel like the BYFPE when you are schooled by someone with more knowledge about a particular subject. The point is, everyone can all still learn regardless of where you are in your career.

I hope you find ***51 Conversations with the Wise Old Fire Protection Engineer*** informative, useful, and entertaining. I welcome your comments, questions, and suggestions for preparing the next fifty-one conversations. Just send them to me at firemarkpress@gmail.com.

Jeffrey Moore
Union, KY
August 2022

About the Author

Jeffrey Moore, PE, FSFPE, CFEI is a practicing fire protection engineer in Union, KY. He has over forty years of experience in fire protection engineering, industrial fire protection, fire protection training, and firefighting. He is a licensed fire protection engineer in multiple states, was selected as a fellow by the Society of Fire Protection Engineers (SFPE), and is a Certified Fire and Explosion Investigator (CFEI). He is also the author of the widely used text *Industrial Fire Control Concepts*, 3rd Edition.

Chapter 1 – The Fire Threat

#1 – Fire Loss Lessons

"Hey Boss," inquired the Brash Young Fire Protection Engineer (BYFPE) of the Wise Old Fire Protection Engineer (WOFPE). "Why aren't there many detailed fire investigation reports for the more recent significant fires. There seems to be a lot of information on major fires occurring before about 1990, but detailed investigative reports on more recent fires are hard to find. These types of reports are a great way to learn from the past, so we don't keep repeating the same mistakes."

"That's a very perceptive observation," replied the WOFPE. "It's true, there is detailed information on earlier fires, but finding credible information, not just speculation, on more recent fires is difficult in many cases."

"Why is that?" asked the BYFPE. "There are important lessons to be learned from all fires."

"Agreed," said the WOFPE. It used to be that within a few weeks of a major industrial fire, everyone in the industry had details about what happened. When I worked in the highly protected risk insurance industry, I remember reading the loss investigation report on a major industrial fire within a couple of months of the fire. And the report came from a competitor because we routinely exchanged educational loss information in those days. If a fire occurred within a particular industry, like the automotive industry, all the auto manufacturers had the information in a short time because they informally shared such information so everyone could learn."

Quizzically, the BYFPE asked, "Why isn't it that way today?"

Letting out a deep breath, the WOFPE replied, "Can you spell L-A-W-Y-E-R-S?"

"Lawyers? What do lawyers have to do with it?" questioned the BYFPE.

"Well, it's not necessarily the lawyers themselves, but the litigious society we live in today," explained the WOFPE. "There used to be multiple investigations of major fires to document the conditions and lessons to be learned. For example, the National Fire Protection Association (NFPA) and the U.S. Fire Administration (USFA) would send a team of investigators to prepare detailed reports and share them with the fire protection community. Today, everybody lawyers up to protect themselves from potential liability as soon as a fire occurs. The last thorough investigative report prepared by NFPA was in the 1990s. All you see today after a major fire is a summary gleaned from news reports and other publicly available information."

"So, nobody is investigating the big fires anymore?" quizzed the BYFPE.

"Oh no, some of the most technically talented fire protection professionals around thoroughly investigate most large fires," explained the WOFPE.

"You mean like the litigation support cases you work on?" asked the BYFPE.

"Exactly," replied the WOFPE. "Most of the fire investigation work I do today is for an attorney or an insurance company representing a potential party in a fire. Just like the manufacturer of the fire alarm system I'm working for on a current case."

"But you said the fire alarm system had nothing to do with the fire. That the fire started in an area of the building that wasn't sprinklered, which is why the building burned down, but the fire alarm system worked," said the BYFPE. "Can't your report be released so others can learn from the fire?"

The WOPFE replied, "Not until all the litigation is over and the case is settled."

"And when will that be?" questioned the BYFPE.

"I'm not sure, but one case I worked on was over ten years from the day I first set foot on the scene for the investigation until resolution of the case in court," explained the WOFPE.

"Ten years!" exclaimed the BYFPE. "It took ten years to get your report released?"

"No," corrected the WOFPE, "My report was never released. The attorneys didn't permit its release because of potential future liability concerns."

"Why?" questioned the BYFPE. "You did an in-depth study of the function of the fire alarm system and found it worked perfectly. So how would the fire alarm manufacturer have any liability?"

"Ah, my young colleague," exclaimed the WOFPE. "You're confused about liability. That's a prevalent affliction. You see, anybody can make a claim of liability against anyone for any reason at any time. It's simply a case of filing the appropriate forms to initiate the legal action."

"That's crazy," said the BYFPE.

"Yes, it seems that way," said the WOFPE, "but that's why much of the technical information on a fire obtained during the investigation by experts never gets released. It's tied up in legal proceedings for years or even decades. In many cases, it never gets out. Thus, we miss the chance to learn from it. Even when we can obtain information on a current fire disaster, we find the same factors present as in many of the past's classic industrial and commercial fires. New lessons aren't much help when we ignore the old ones."

The BYFPE responded, "I still don't see why a company that doesn't have anything to do with a fire should be concerned about liability. If they aren't at fault, why should they have any liability."

The WOFPE smiled and said, "I would agree, but that's not how our civil legal system works. As an attorney friend explained it to me, when you look up liability in the dictionary, it's defined as an exotic game played by two teams of attorneys in front of a field of twelve referees whose job it is to determine which participants in a litigation matter have the deepest pockets. The winner of the game and assignment of liability often has nothing to do with actual causal factors."

The BYFPE looked at the WOFPE quizzically and asked, "You're friends with an attorney?"

#2 – Fire Protection Success Stories

"Hey, Boss," hailed the Brash Young Fire Protection Engineer (BYFPE) to the Wise Old Fire Protection Engineer (WOFPE). "I finished reading all those case histories you gave me, and I understand how important it is to learn the lessons from those losses, but if you just relied on those case studies, you'd get the idea that we can't seem to do anything right. Don't you have any case histories that are success stories instead of all these disasters?"

"Success stories? You want success stories?" quizzed the WOFPE as he pulled a folder out of his desk and opened it to reveal several clippings from newspapers and magazines. "Come and sit. I'll give you some success stories."

Pulling out some of the clippings, he handed one to the BYFPE. "Here's one. Read it out loud."

The BYFPE started, "The Mt. Haven Fire Department responded to a reported fire at the United Chemical Company plant on Iberia Street last night at 10:42 PM. On arrival, they found the facility evacuated with a small fire in the waste handling system contained by the building sprinkler system. A spokesperson for United Chemical credited the sprinkler system with containing the fire and said there were no injuries and no disruption of production operations. No further details were available."

"Here's another," said the WOFPE handing the BYFPE another clipping. "What does it say?"

"Well, the headline is, FIREFIGHTERS CREDIT SPRINKLERS WITH CONTROLLING FIRE AT RESEARCH LABORATORY," said the BYFPE. Then, perusing the article, he continued, ". . . Battalion Chief Jack McCabe said the fire started in an employee break room, but a single sprinkler quickly extinguished the fire. Damage was minimal, and there were no injuries . . . "

"Pretty boring stuff, don't you think?" probed the WOFPE. "Check this one out," he said, handing the BYFPE another clipping.

The BYFPE looked at it and started, "This one says SOLVENT FIRE SUPPRESSED BY SPECIAL FIRE EXTINGUISHING GAS. A specialized gas firefighting system extinguished a Tuesday afternoon fire at the

Acme Ajax plant involving volatile chemicals. The city fire marshal reported that a worker dropped a drum of flammable solvent from a forklift in a specialized storage room at the Acme Ajax Broad Street facility, causing it to leak and create highly volatile vapors that ignited. The burning material released a specialized gas that extinguished the fire." The BYFPE paused and looked quizzically at the WOFPE. "What specialized gas are they talking about?"

"There was no specialized gas," explained the WOFPE. "It was a carbon dioxide extinguishing system installed about twenty-five years ago. Properly maintained, when called upon to work, it did. But unfortunately, even the catchy wording of the news writer couldn't kick up the excitement level enough to be front-page news. So this clipping was just a filler used on page B-7."

"That's the thing about our success stories," explained the WOFPE. "They're boring. When everything works correctly, there is no news. There are no breathtaking photos. There's no dramatic video. There is no 300-page loss investigation report. There are no presentations at national conferences. Nobody but those of us in the fire protection field even cares."

He continued, "We have success stories every day. I even suspect they are vastly underreported. Think about it. Every day there are small fires at industrial facilities like Amalgamated Industries, United Commercial Chemicals, or similar places. They dispatch the plant fire brigade, find that a sprinkler fused and controlled the fire, restore the sprinkler system, and everything goes back to normal. Or a trained employee uses a fire extinguisher to extinguish a trash fire before it spreads. There is generally no report to the public fire department or the insurance company because any damage is well below the policy deductible. None of those success stories ever get included in the National Fire Incident Reporting System from which most of the published fire statistics come. Our successes are boring and don't make exciting reading for most people.

Chapter 2 – The Science of Fire

#3 – Fire Development Myths and Misconceptions

The Wise Old Fire Protection Engineer (WOFPE) was at his desk engrossed in the drawings and design analysis for a new building. He had a stack of codebooks open and was checking various references with a scowl on his face when the Brash Young Fire Protection Engineer (BYFPE) broke his concentration with, "Hey Boss."

"What is it?" snapped the WOFPE as he scribbled a note on his tablet.

"It's OK," uttered the BYFPE sheepishly. "I didn't mean to interrupt you. I can come back later."

"No, no, no," said the WOFPE as he waived his protégé into one of the chairs across from his desk. "The only thing you interrupted is my rage at this idiot designer who's apparently using a codebook from about 1950. Either that, or he is an actual idiot. So, sit down and tell me what's on your mind?"

"Have you ever heard of the 10-minute rule in firefighting?" said the BYFPE settling into a chair opposite the WOFPE.

The WOFPE looked at him pensively and responded, "Where did you hear about the 10-minute rule?"

"I was at that old mill building used by the Acme Ajax Corporation to clean and recycle metal 55-gallon drums. One part of the building is an addition added about fifty years ago, and they still don't have sprinklers in it. And it's all wood construction. The plant manager said he asked the fire department about sprinklers and was told they weren't required unless they make changes to the building and bring it up to the current code. He said the fire inspector also said the building meets the 10-minute rule, so he didn't think sprinklers were necessary. The inspector said it would be a big expense for nothing because their first due engine company was only about five minutes away. It floored me that the fire department said sprinklers would be an unnecessary expense. What the heck is he talking about with the 10-minute rule stuff?" questioned the BYFPE.

The WOFPE smiled and said, "I would venture a guess without meeting the gentleman from the fire department that he is near retirement, hopefully, and is probably counting the days until he can start his pension. He's probably one of those individuals with one year of experience many times over. I say that because, like you, I can't imagine a modern, fully informed fire service professional saying sprinklers are an unnecessary expense."

"But what about this 10-minute rule stuff?" voiced the BYFPE.

The WOFPE pushed his chair back and put his feet up on the desk knocking one of the codebooks onto the floor. He leaned his chair back and began, "Even when I started in the fire service, and I don't yet consider myself old, I had a couple of fire training courses where they talked about the 10-minute rule. It said you have about ten minutes from when a fire breaks out until it flashes over and spreads. A hundred years ago, unless you were rich, you probably didn't have a lot of fancy furniture and other furnishings stuffed into your house. And we hadn't yet started using plastics and synthetics in everything. Based on the combustible loading common for that era and anecdotal evidence, it was common to believe if you could suppress a fire in the first ten minutes, you could save the building. And I emphasize that it was anecdotal evidence at best. Even though there was no fire testing to prove it, this idea made its way into some training programs for firefighters."

"That's crazy," exclaimed the BYFPE. "The latest fire testing shows we can reach flashover in a typical residential room in less than four minutes and sometimes within two minutes. So if you wait ten minutes to attack the fire, I can almost guarantee you'll lose the building. I even saw testing in typical office settings that showed the same fire behavior–flashover in three or four minutes. So why would they be teaching firefighters they have ten minutes?"

"They don't teach firefighters that and haven't for a long, long time," explained the WOFPE. "Unfortunately, I think you've stumbled onto one of the dinosaurs that haven't yet moved into the current century. If I were you, I would call the plant manager and explain the ten-minute rule's fallacy and why sprinklers are essential if he wants to save the building. I'm going to call the fire chief and just casually mention this incident to ensure it's an isolated one."

Chapter 3 – The Fire Control System

#4 – Insurance Loss Control Surveys

"Hey Boss," called the brash young fire protection engineer (BYFPE) to the Wise Old Fire Protection Engineer (WOFPE), "Our client decided they weren't going to spend the money to upgrade the sprinklers and make the other improvements we recommended for that remote warehouse. They said it must not be much of a problem since their property insurance company isn't pushing it. I didn't know what to say since that sounded logical."

"Well, sit down and let's talk a bit about property insurance companies and how they operate," counseled the WOFPE.

"First, a property insurance company is in business to make money. Anything they do or don't do is for their own best interests," explained the WOFPE.

"But Boss," interrupted the BYFPE, "insurance companies advertise that they help their customers limit risks, solve fire protection and loss control problems, and look out for their bottom line. All those insurance inspectors marching through facilities, writing up potential code violations, and documenting fire and explosion hazards have to be helping their customers identify and reduce the risk of a loss at least some. So how can you say they are only looking out for themselves?"

The WOFPE looked at his young protégé and sighed. "You're not listening. I didn't say their efforts couldn't help their clients, but do you think an insurance company will spend the time, money, and personnel resources to 'help' their customers without benefitting themselves?

"Of course not," replied the BYFPE. "But ..."

The WOFPE interrupted, "A thorough survey of a facility conducted by an experienced and knowledgeable insurance company loss control representative can identify conditions affecting the probability of a fire or explosion, and the accompanying property damage and operational disruption likely to follow such an event. When conveyed to and acted upon by the client, this information enables the client to handle the risk better. However, what is the real reason for that insurance loss control representative making a "free" loss control survey of the facility?"

"You sound skeptical of their motives," said the BYFPE. "That surprises me since you spent many years doing those loss control surveys and training other engineers to conduct them."

The WOFPE sighed and started again. "You are correct. I spent many years filling those roles. However, I never lost sight of the purpose of a loss control survey."

Said the exasperated BYFPE, "So, if it's not to help the client, then what is the purpose?"

The WOFPE explained, "The primary purpose of an insurance company survey is to gather information for the insurance company underwriters so they can profitably underwrite the risk. The underwriters need information on the type of buildings, their construction, the operations, the adequacy of the fire protection systems and features, and, most importantly, the attitude of the facility management to establish the premium or price of the insurance and make a profit. An important part of that report is documenting the deficiencies by developing recommendations for improvements. These recommendations are how the loss control representative conveys to the underwriter the deficiencies found at the facility."

The BYFPE looked unconvinced. "Well, if that is the real purpose of the insurance company surveys, why does the insurance company share all of that information with the client in a loss control or loss prevention survey report with recommendations?"

The WOFPE slowly shook his head. "The insurance company doesn't share all the information they gather during their surveys with the client. Most reports provided to a client give only a general description of the risk and a list of the recommendations that identify deficiencies and the means to correct them. The report provides a wise facility manager with a tool to evaluate the fire risk within the facility. Still, it is by no means an all-inclusive list of every possible hazard or problem. Just because the insurance company doesn't mention a problem doesn't mean it's not a problem. Or possibly the problem wasn't evident during the brief time the insurance company engineer was in the facility. The loss prevention engineer is there once or twice a year. Effective loss prevention and control is a 24/7/365 proposition that facility management must embrace."

#5 - More About Insurance Loss Control Surveys

"Boss, I'm confused," declared the Brash Young Fire Protection Engineer (BYFPE). "Earlier, you told me that large property insurers sent their engineers through their insured risks to identify loss potential so they can underwrite the risk at a profit."

"Yes," said the Wise Old Fire Protection Engineer (WOFPE), "and your question is?"

The BYFPE continued, "Well, the best thing I could say about the facility I just visited is that it wasn't on fire when I left."

"And your question is?" repeated the WOFPE.

"The plant manager said Industrial Mutual insures the facility, and they just had an insurance inspection. He showed me the report. It was clean. There wasn't a single recommendation on it. The plant manager said that given the reputation of Industrial Mutual, if they didn't have a problem with the facility, he was satisfied that things were in pretty good shape."

"And your question is?" repeated the WOFPE with a touch of impatience.

"From what I saw in the plant, there is no way even a first-year loss prevention associate could have missed the holes in the firewall, broken fire doors, sprinkler systems out of service, faulty alarm devices, flammable liquids stored in the manufacturing area, and I could go on and on . . ."

"Stop!" commanded the WOFPE. "For the fourth time now, what is your question?"

"How could Industrial Mutual have missed all this stuff during their survey? Why was the report clean of any recommendations? And finally, how could an HPR (highly protected risk) insurance company insure such a dump?"

Slowly exhaling, the WOFPE, now in a calmer state, said, "Well, my eager colleague, let's take those questions one at a time. First, I doubt an experienced loss control engineer 'missed' all the things you describe."

"But the report is clean. No recommendations. I saw it," protested the BYFPE.

"You saw the report provided to the client. However, you didn't see the complete report intended for internal use by the insurance company. That complete report likely contains a list of the deficiencies you described and others you might not have seen. This list of deficiencies and any recommendations to correct them probably didn't make it to the plant manager for any number of reasons."

"Like what?" questioned the BYFPE. "Based on our earlier discussions, you said the insurance company wants the client to have that list so they can correct the deficiencies and lower the risk to the insurer."

"True enough," explained the WOFPE, "but what if the value of the entire facility is less than the insurance policy deductible? You said the facility was a subsidiary of Acme Ajax International, which I know from other work, has a property insurance policy deductible of around $15 million. Is this facility and everything in it worth more than $15 million?"

"Probably not. Based on per square foot costs, it's probably worth substantially less than that," replied the BYFPE with a dawning look of comprehension. "Then the insurance company probably isn't too worried about a loss at this facility, so they aren't going to push for improvements, are they?"

"Exactly," confirmed the WOFPE. "It could also be that at the corporate level, the client knows the facility is a dog but has plans to be out of it in the next five years. So, they tell the insurance company there is no sense trying to beat them up about a facility they know is a dog, are trying to unload it, and won't be investing any money there. The insurance company agrees, buys some reinsurance to protect themselves, and places all the recommendations on an abeyance list, so the next engineer doesn't waste time making the same recommendations again. After all, as long as they don't have to pay out for a large loss, they aren't going to expend resources trying to change the situation."

"I see," said the BYFPE. "That answers my questions about how an experienced engineer could go through the facility and there not be any recommendations on the report, but it doesn't answer my question about how a large, well respected HPR insurance company could insure a dump like this."

The WOFPE smiled and explained, "Your newfound understanding of property insurance loss control services and reports impresses me. To address your last question, remember that all property insurance companies, and insurance companies in general, are in business to make a profit. Regardless of what they advertise, never confuse insurance with protection. You can buy insurance for anything if you are willing to pay the premium the underwriter demands to assure the chance of a reasonable profit."

"I've even heard of a building owner buying extra liability insurance for a multiple fatality fire after the fire occurred," said the WOFPE. "The building owner was actually buying bulk legal services from the insurance company to litigate the claims arising from the fire. The insurance company estimated what it would cost in damages to the claimants, the legal fees to litigate the case, and added a contingency to assure when everything was said and done, they still ended up with a profit."

"Never lose sight of the fact that insurance companies are in the business of buying risk. For a payment, your insurance premium, they buy a portion of the risk of loss. If they do a good job of underwriting and underwrite a sufficient number of risks, they end up with a profit at the end of the year. As a wise old mentor once told me, the problem with insurance underwriting is that too often, the perfume of the premium overcomes the stench of the risk."

INDUSTRIAL MUTUAL
PROPERTY LOSS CONTROL SURVEY
CONFIDENTIAL

ACME AJAX AMALGAMATED CORP. PLANT 42 10245 IBERIA ROAD MOUNT GILEAD (MORROW COUNTY), OH USA	**DATE:** November 14, 2021 **BY:** R.E. Beard **LOC ID:** 122315/111456

PROPERTY DESCRIPTION

This location has two modern detached industrial buildings that together are called Plant 42. One building is a single-story, 75,000 sq.ft., light noncombustible building with metal panel walls and a built-up metal deck room supported on an unprotected steel frame manufacturing building producing widgets of various sizes for use in gonkulators assembled at other facilities. The warehouse is a single-story, 120,000 sq.ft., light noncombustible building with metal panel walls and a built-up metal deck room supported on an unprotected steel frame. The site also includes a single-story, 3500 sq.ft. metal storage building for landscaping maintenance equipment.

Plant 42 provides feed materials to three other Acme Ajax Amalgamated Corporation facility assembly gonkulators for the commercial and military markets.

6 – Understanding Fire Testing

"Tell me," said the Wise Old Fire Protection Engineer (WOFPE) to the Brash Young Fire Protection Engineer (BYFPE). "The first draft of your report on the Acme-Ajax Corporation's new anechoic test chamber had a recommendation for sprinklers inside the chamber. So why doesn't the current version of your report recommend sprinklers?"

"Easy," said the BYFPE. "They showed me that the material lining the chamber wouldn't burn."

"It won't burn?" questioned the WOFPE. "It's made of carbon-impregnated, expanded polyurethane. So how did you decide that it won't burn?

"They gave me a two-foot-high cone of anechoic liner material and defied me to set it on fire. So I took it out back and used a match, a lighter, and even a propane torch to try to set it afire and couldn't. It would flame a little when the torch flame impinged directly on the material, but as soon as I removed the torch, the flame went out."

"Really," said the WOFPE. "You couldn't set it afire, so your conclusion is it won't burn, and hence sprinklers are not needed?" He paused and said, "Let's adjourn outback where you have this material. And bring your propane torch as well."

"See," said the BYFPE as he directed the torch flame against the side of the material and removed it. "It flames when the torch is on it but self-extinguishes as soon as I withdraw the torch."

"Well, I'll be darned," said the WOFPE, "Let me see that torch for a minute and hand me that piece of two-by-four laying over there." The WOFPE directed the torch flame against the wood two-by-four and watched as the wood began to flame around the point of flame contact. Then, he removed the torch and watched as the flame immediately went out. "Look at that! In all my years, I never realized that wood two-by-fours don't burn. Think of all the buildings provided with sprinklers just because they are of wood construction.

He dropped the two-by-four and said, "Hand me that scrap of plywood over there." Holding the torch flame against the plywood and then removing it, he again watched as the flame on the plywood extinguished.

"Now I feel really foolish," said the WOFPE. "All these years working in fire protection, and I never knew that wood two-by-fours and plywood won't burn. I feel so stupid. Have you known this all along?"

Sheepishly, the BYFPE said, "Of course, two-by-fours and plywood burn. They burn quite well but need exposure to a higher energy heat source before igniting and continuing to burn. And I get your point. A small-scale, ad hoc test doesn't accurately represent a material's burning characteristics in the form and configuration in which we might use it. I had better go look up the full-scale fire test results of the anechoic material before I say sprinklers aren't needed."

"That would be a wise move," said the WOFPE. "Radiant heat feedback from surrounding materials and surfaces plays a vital role in determining a material's burning characteristics in actual fire situations. Therefore, always conduct testing in the same configuration and under the same conditions in which the material is used in actual practice. And always be skeptical of small-scale, ad hoc testing conducted without the benefit of a proven fire test procedure."

#7 – Don't worry, the fire department is only five minutes away.

"I need some help responding to some design review comments," said the Brash Young Fire Protection Engineer (BYFPE) to the Wise Old Fire Protection Engineer (WOFPE). "With the agreement of the project architect, the project manager for the new Amalgamated Industries Office Building says that they only need the absolute minimum fire protection because the fire department is less than five minutes away."

"Ah, and age-old misconception," observed the WOFPE. "The distance between the fire department and the facility has nothing to do with the level of fire protection needed for a building or operation."

"That's what I told them during the design review meeting," explained the BYFPE. "But I didn't make much headway. I tried to explain the fire life cycle timeline described in "**Industrial Fire Control Concepts**," but nobody was very receptive to my ideas."

"A meeting with the fire department might move the design team off of dead center," offered the WOFPE. "I had a similar situation a few months ago, and the city fire marshal gave a presentation to the client on the First Interstate Bank fire in Los Angeles to demonstrate the fallacy of equating fire department proximity to fire protection."

"I'm familiar with the fire," said the BYFPE, "but how does a high-rise fire in mid-town Los Angeles equate to an industrial facility in our area?"

"An explanation of the First Interstate Bank provides the public with a glimpse at actual fire department response time and what it means," explained the WOFPE. "If you remember, the building was a 62-story high-rise building in mid-town Los Angeles. The Los Angeles City Fire Department, one of the best-trained, best-equipped fire departments in the world, received an alarm for a fire on the 12th floor of the building at 10:37 PM. The first fire company arrived two minutes later at 10:39 PM to find a fire involving most of the 12th floor. It then took until 11:15 PM to move enough firefighters and equipment up in the building to mount an attack on the fire. In the meantime, the fire spread upward to involve the 13th floor. It was 2:19 AM the following day before the incident commander declared the fire under control."

The WOFPE continued, "The effort involved the response of 64 fire companies and 383 firefighters. Unfortunately, the quick arrival of the fire department did not equate to quick fire control. In this case, a well-trained, well-equipped fire department arrived within 2-minutes, but it still required over 3.5-hours to establish control of the fire. The extended time between the arrival of the fire department and actual control of the fire was due to the time necessary to assemble sufficient personnel to attack the fire with a rate of flow adequate to control it and the time to provide the logistical support needed to mount a sustained fire attack. The critical flow rate, or flow rate necessary for fire control, is an important concept. Many people measure facility "fire protection" based on how fast the fire department can get to the front door. Unfortunately, the public does not understand that quick fire department response does not equal quick fire control."

"Outside of those within the public fire service, the planning and logistics required for an effective fire attack in a large commercial or industrial facility and the time necessary to mount the attack are not well understood. In reality, there is much more to effective manual fire suppression operations in industrial, commercial, and institutional facilities than merely 'putting the wet stuff on the red stuff.' While quick, efficient response by the fire department is an essential element of the total fire control system, proximity to the fire department does not equal good fire protection."

"How do we apply a high-rise building case study to an industrial facility?" questioned the BYFPE

"A large area industrial facility is just a high-rise building laid on its side," explained the WOFPE. "A fire truck pulling up to the front gate in two minutes does not suppress the fire. It may be ten, fifteen, thirty minutes or more between the arrival of the first fire apparatus and firefighters at the facility and the establishment of the water supply and logistical systems necessary to apply the critical rate of water flow to bring the fire under control. That's why we need the sprinklers, firewalls, fire barriers, and other fire protection features in the facility. They can limit fire spread and establish fire control while the fire department sets up to extinguish the fire."

#8 – Don't Worry, It Meets the Code.

"Hey, Boss," hailed the Brash Young Fire Protection Engineer (BYFPE) to the Wise Old Fire Protection Engineer (WOFPE). "Can we discuss your comments on my fire protection design brief for the Acme Ajax Distribution Center renovation project?"

"Sure, let's talk," said the WOFPE on the way into his office. "What specifically did you want to discuss?" he said as he sat behind the desk.

"Every place I referred to the 'minimum requirements of the applicable code,' you put a big question mark," explained the BYFPE. "I don't understand your question. Aren't we supposed to always meet minimum code requirements?"

Taking a marked-up copy of the design brief from the BYFPE, the WOFPE responded, "Of course, we must meet the minimum code requirements and your design brief explains compliance with the minimum code requirements. What it doesn't explain is how you are meeting the site-specific needs of the facility based on their operating conditions and fire risk management objectives."

"Um, I'm not sure what you mean," admitted the BYFPE. "The project architect is driving every discipline to keep costs as low as possible to meet the project budget. So I'm just providing them with the minimum level of fire protection specified by code."

"I understand that," explained the WOFPE, "but does that code mandated minimum level of protection meet the client's needs in terms of property protection and protection of the facility's mission or ability to continue business operations?

The BYFPE responded, "Well, both the client project manager and the architect are happy with the brief because it looks like the project will come in under budget."

The WOFPE sighed and questioned, "Do they understand that meeting just the minimum code requirements doesn't mean their building or continued business operations are protected? Do they understand that buildings meeting minimum code requirements burn down every day? Does the client project manager understand that the day after they get the certificate of occupancy that their building could burn to the ground, and they might be out of business?"

"What do you mean?" asked the BYFPE plaintively. "If it meets the public fire and building codes, it should be protected."

"No!" exclaimed the WOFPE. "If it meets the PUBLIC fire and building codes, the PUBLIC is protected. The PUBLIC fire and building codes are not written to protect the building and its operations. The PUBLIC codes protect the PUBLIC from the building and its operations. If the building catches fire and burns completely to the ground, but everyone gets out alive and unhurt, and the fire doesn't spread beyond the property line, every one of the intended fire safety objectives of the PUBLIC codes is met. They protected the PUBLIC from the building. The PUBLIC codes do not address and do not care whether Acme Ajax still has a facility after the fire or whether they are still in business. That little fact often escapes the understanding of most property owners and building designers. Meeting the minimum requirements of the PUBLIC codes does not assure protection of the building or its operations."

"Well, what about the NFPA codes and standards," said the BYFPE defensively.

"Again, those documents are minimums developed by a committee with varied interests that may or may not align with the best interests of Acme Ajax," explained the WOFPE. "The best way I can describe it is with a pizza."

"A pizza?" asked the BYFPE incredulously. "How can you explain the NFPA codes and standards with a pizza."

"Well, let's assume," started the WOFPE, "That we are part of a new NFPA technical committee made up of all different interest groups and that our task is to develop the specifications for the pizzas we are going to eat for dinner tonight. The way the NFPA process works is that our committee of twenty to thirty people must come to a two-thirds consensus on the requirements, meaning that two-thirds of our committee members must agree on something before it becomes part of this new standard on pizzas."

The WOFPE continued, "Now, as our committee begins to deliberate, there are competing interests. For example, we might have a few committee members that don't like green peppers and onions on their pizza because they get heartburn. There might be some members that are lactose intolerant, so they are going to lobby against extra cheese.

Another group might not like anchovies, so they go to the anti-green pepper and onion group and say, if you vote against anchovies, we'll vote against green peppers and onions. These discussions and deliberations continue throughout the day until we reach a two-thirds consensus of the committee on what type of pizzas to order for dinner. Any idea what kind of pizza we'll probably eat for dinner tonight?"

"Not a clue," answered the BYFPE.

"Well, assuming we can reach a consensus," explained the WOFPE, "we'll probably be eating plain cheese pizza because, with all the different ideas of the different committee members as to what should be on the pizza we order, the lowest common denominator we can probably agree on is plain cheese pizza."

"Just plain cheese?" questioned the BYFPE. "Doesn't sound like much of a dinner to me."

"It doesn't to me either," agreed the WOFPE, "but it's the consensus we could reach. I think that a super deluxe, meat lovers' pizza with extra cheese is a much better option, and I can order that for myself if I want. But the minimum standard is plain cheese pizza. You need to look at every NFPA standard and every other document developed through the consensus committee process as a plain cheese pizza. It's not necessarily the best; it's just the minimum. If it doesn't meet our needs, we are free to go above and beyond the minimum."

The BYFPE pondered that statement for a moment and said, "I understand, but how does this apply to my design brief?"

"You need to make sure the client understands that compliance with the minimum requirements of the code, wherever it comes from, is like a plain cheese pizza," explained the WOFPE. "But based on the site-specific needs and conditions of the facility, they might need to go beyond the minimum to assure meeting their site-specific needs and objectives."

The WOFPE continued, "For example, if you look in the PUBLIC codes or the applicable NFPA standards for paint spray booths, you'll find that they require 'a' fire suppression system. It might be a sprinkler system, a dry-chemical extinguishing system, or some other approved extinguishing system, but the minimum code requirement is to provide a single system. If that is the case, why do some automotive assembly facilities have three levels of fire suppression in their paint spray booths?"

"Because their need is not only to detect and suppress a fire but to do it before the fire causes sufficient damage to shut the process down, volunteered the BYFPE with confidence. They know they aren't selling any cars without paint."

"Exactly!" answered the WOFPE. "Before you submit your design brief, you need to talk to the client to find out exactly what their fire risk management objectives and loss tolerance are for the project. If their only objective is to obtain a certificate of occupancy, compliance with the minimum requirements of the applicable codes is probably adequate. However, like the automotive assembly facility, I suspect they have internal fire risk management goals and objectives that require a level of protection exceeding the minimums of the PUBLIC codes and applicable NFPA standards. That's what you need to explain in your design brief."

FOR
FIRE
OPEN
THEN PULL DOWN HOOK
3334

Chapter 4 – Fire Alarm Systems

#9 – Umbrella Systems

"Hey Boss," said the Brash Young Fire Protection Engineer (BYFPE) as he gently rapped on the doorframe to the office of the Wise Old Fire Protection Engineer (WOFPE). "Got a minute?"

"Sure," said the WOFPE looking up from a file and motioning the BYFPE toward a chair across from his desk. "What's on your mind?"

"I was at the design kickoff meeting for the new fire alarm system project for International Amalgamated Industries this morning," explained the BYFPE. "The plant engineer kept telling everyone this would be an 'umbrella system' and that it should be designed as an 'extension of management.' Later, I talked to her during a break, and she said to thank you because she got those ideas from you. I didn't want to seem out of the loop, so I didn't ask her about it, but would you mind telling me what the heck she was talking about?"

"I'll be darned," chuckled the WOFPE leaning back with a satisfied grin on his face and clasping his hands behind his head. "She was listening."

"She was listening to what?" questioned the BYFPE.

The WOFPE smiled while staring at the ceiling. Then, after a moment, he dropped his hands, leaned forward in his chair, and moved a couple of files from his desk to the out basket. "It's probably been ten years or so now, but once upon a time, she was in a fire alarm class I taught. She had just started in the Facilities Engineering Department at Amalgamated, and they sent her to a course I was teaching to give her some background in fire alarm systems. At that point, she was only a year or so out of engineering school. I remember she sat near the rear of the class. From what I remember, she seemed interested, but I don't think she asked or answered a question during the whole five days of the program."

"How do you know it was her?" quizzed the BYFPE.

"Oh, it was her," said the WOFPE confidently. "Ole' Mike McGinnis was the plant engineer at Amalgamated at the time. He called and said he was sending her. Mike wanted me to keep an eye on her and let him know how she did. He said she seemed pretty sharp and that from what he had seen so far, he joked she'd probably be taking his job if he didn't watch it. Over the years, she steadily moved up until she took over as the Plant Engineer when Mike retired last year."

"I can see you enjoy your reminiscence of the good old days," cracked the BYFPE drawing out old for emphasis. "But, what does that have to do with my question about umbrella systems and the fire alarm system being an extension of management?"

"Sorry, I know your attention span is short, so let me get to the point before you have to check your phone three more times," responded the WOFPE with a touch of sarcasm.

The BYFPE put his phone in his pocket and looked intently at the WOFPE. "Sorry," he said.

"Back then, I introduced fire alarm systems, explaining that if properly designed and installed, a facility fire alarm system is an extension of facility management. After all, facility management is responsible for overseeing and controlling everything in the facility," explained the WOFPE. "When properly designed, the fire alarm system watches over and monitors the operation of all other fire protection and life safety features in the facility. For example, it monitors the sprinkler systems, fire pumps, and special extinguishing systems, and today can even track the portable fire extinguishers to confirm they are in place and fully charged. There are many other features and functions that fire alarm systems can monitor and control. For example, it monitors and controls the elevators during an emergency. We use the fire alarm system to shut down the HVAC systems and start smoke exhaust and smoke control systems. We can even monitor critical industrial processes. For example, over at the Acme Ajax plant, the fire alarm system monitors the pH level of the effluent from their water treatment facility. If it detects a reading out of normal range, it sends a supervisory signal to the fire alarm system. Security then dispatches someone from the environmental engineering group to check it out. That's the idea of a fire alarm system serving as an extension of management."

"Oh, now I see," said the BYFPE as he settled back in his chair. "The idea of the fire alarm system as an 'umbrella system' is just another way of saying that the fire alarm system oversees all the other fire protection systems in the facility."

"Exactly," said the WOFPE smiling satisfyingly at the BYFPE. "A fire alarm system is only one part of a facility's overall 'fire protection' system, but it's an important part. It must meet the site-specific needs and objectives of the facility it serves so it can function as an extension of a concerned management's control of the facility."

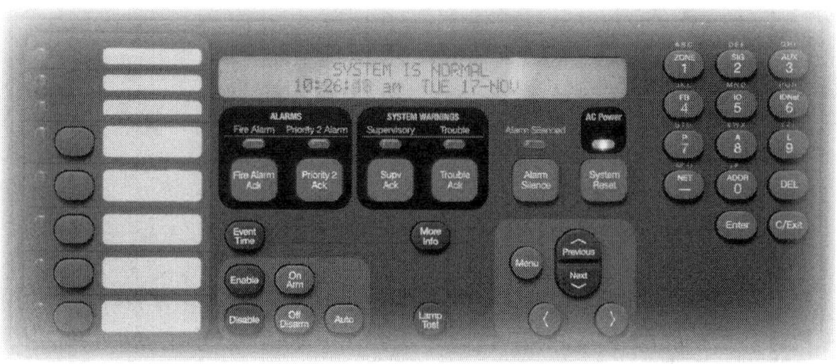

#10 – Heat Detector Listed Spacing

"Hey, Boss," hailed the Brash Young Fire Protection Engineer (BYFPE) to the Wise Old Fire Protection Engineer (WOFPE). "Can you explain why UL uses sprinklers to determine the listed spacing for a heat detector? The test seems to assure the heat detectors respond as fast or faster than one particular type of standard sprinkler. But what about all the other makes and models of sprinklers? What about quick-response sprinklers? The test method just doesn't make any sense to me."

"Let me give you a little of the history around the UL test for heat detector spacing," said the WOFPE.

"The current UL standard for heat detectors is UL 521, "Standard for Heat Detectors for Fire Protective Signaling Systems." The first edition of this standard dates to 1961. Unfortunately, the standard doesn't explain why they use sprinklers to establish the maximum listed spacing for a heat detector. However, a review of earlier fire protection references sheds some light on why they might compare heat detector response to sprinklers.

"For example, an 1896 edition of the "Handbook of the Underwriters' Bureau of New England," which was essentially the first edition of the "NFPA Fire Protection Handbook," and the 1899 Transactions at the Third Annual Meeting of the NFPA in Boston, both mention the use of heat detectors in buildings protected by automatic sprinklers. Although they referred to the heat detectors as 'thermostats,' there was much discussion about determining the sensitivity of the heat detectors relative to the sprinklers. Property insurance underwriters reported experiencing excessive losses when the sprinklers operated before the heat detectors. The sprinklers cooled the environment preventing activation of the heat detectors and sounding the 'electric alarm.' The NFPA document establishes maximum spacing with the stated intent of assuring the heat detectors always operated first so the alarm would sound. Without an alarm, insurance companies worried about 'unnecessarily large water damage' as the sprinklers operated without alerting anyone. This indicates that the initial testing of heat detectors against sprinklers intended to assure the heat detectors always operated first."

"They used heat detectors and sprinklers?" questioned the BYFPE.

The WOFPE continued, "Until the common use of alarm valves using water to sound a mechanical bell, heat detectors were the only means of sounding an alarm. They didn't have vane-type or pressure switch-type sprinkler waterflow alarms in those early days of sprinklers."

"So, we still test heat detectors against sprinklers because that's what we've always done?" asked the BYFPE incredulously.

"That probably has some basis in the roots of the current test method," explained the WOFPE, "but it appears UL maintains the current test protocol because it provides a consistent means of comparing the performance of various heat detectors against a common standard which is the specific make, model, and lot number of the sprinklers used for the test."

"That doesn't sound very scientific," mumbled the BYFPE.

The WOFPE expounded, "The current UL test standard contains an array of environmental and electrical tests to confirm performance under various conditions. The fire test to determine the listed spacing is only one of the many tests required for UL listing. The continued use of a fire test using a specific sprinkler from a particular manufacturer's lot provides a consistent means of comparing the response of one heat detector to another. That's probably why they still use sprinklers to determine the listed spacing of heat detectors. After all, every sprinkler is also a fixed temperature heat detector.

#11 – Heat Detectors "Not a Life Safety Device"

"Hey, Boss," hailed the Brash Young Fire Protection Engineer (BYFPE) to the Wise Old Fire Protection Engineer (WOFPE). "Why is every heat detector labeled with NOT A LIFE SAFETY DEVICE or similar wording?"

"Lawyers," replied the WOFPE.

"Oh, come on," exclaimed the BYFPE. "What do lawyers have to do with the labeling on heat detectors?"

"Well, the answer is lawyers, but I guess I need to give you the full story," explained the WOFPE.

He continued, "Hopefully, this doesn't happen today, but at one time, there were unscrupulous, uninformed, untrained individuals, we called them 'trunk slammers,' selling and installing heat detectors as cheap home fire protection devices. Of course, as you know, a typical heat detector will not respond fast enough to sound the alarm, awaken sleeping occupants, and provide the warning necessary to escape before fire conditions are lethal. A fire sufficiently large to actuate a heat detector is likely to be large enough to be producing dangerous levels of carbon monoxide and other toxic gases. Consequently, heat detectors are not considered suitable for applications where life-safety of the occupants is the primary objective."

"Anybody should know that," agreed the BYFPE.

"At one time, whether it was because they didn't care or they were just ignorant of the facts, there were companies that did just that," explained the WOFPE. "Notwithstanding the fact that DE-tection is not PRO-tection, we lost some people in fires where somebody installed heat detectors in a misguided attempt to protect the occupants. The ensuing lawsuits put all the unscrupulous fire alarm companies on notice that if you try to sell heat detectors to homeowners to protect the occupants, you will pay when someone is injured or dies."

"To keep anyone from misusing their equipment, manufacturers started labeling all of their heat detectors 'NOT A LIFE SAFETY DEVICE' so no one would get the idea they are a substitute for a smoke detector or a smoke alarm."

The BYFPE paused for a moment and then said, "So it wasn't a problem with the heat detectors, but the idiots," he paused, "excuse me, I mean the completely uninformed individuals who know nothing about fire protection and fire science, misusing the equipment for an application for which it was never intended."

"Exactly," said the WOFPE nodding his head affirmatively. "We don't want anyone put at risk by believing that a heat detector is equivalent to a smoke detector or smoke alarm."

#12 – Fire DE-tection versus Fire PRO-tection

"Hey Boss," called the Brash Young Fire Protection Engineer (BYFPE) to the Wise Old Fire Protection Engineer (WOFPE) at the other end of the conference room as the meeting ended and people filed out.

"The plant manager from Acme Ajax emailed me about a fire alarm system they are buying for that new building they just leased. I'll forward the email with the details to you."

"Fire alarm system?" questioned the WOFPE as he crossed the room and sat down across from the BYFPE. "When we surveyed the facility for their due diligence inspection, we told them the sprinkler system was deficient, and they should upgrade the sprinklers and water supply before they moved any rack storage in. The alarm system was fine. What gives?"

"Well, the plant manager said the building owner balked at the cost of a new sprinkler system and possibly a fire pump but did agree to put in a brand new fire alarm system with smoke detection installed throughout. He said the building owner told him that the fire alarm company told him it would improve the protection in the warehouse at a fraction of the cost of a new sprinkler system," explained the BYFPE. "And, he liked the idea of getting an early detection system."

Did you explain to him that there's a significant difference between fire DE-tection and fire PRO-tection?" questioned the WOFPE.

"The fire alarm sales engineer said he had done this very same thing in several locations and would even prepare all the paperwork for the Fire Marshal to get it approved as 'equivalent' protection," the BYFPE responded defensively.

"What?" exclaimed the WOFPE leaning toward the BYFPE. "It's not equivalent, and it's not PRO-tection. As far as being an equivalency, it's the same as your doctor telling you that taking an aspirin for a brain tumor is the same as surgery to remove the tumor. Likewise, a sprinkler system and a fire detection system have two different purposes. You can't substitute one for the other."

"But... but..." protested the BYFPE.

Curtly interrupting, the WOFPE continued, "But nothing! DE-tection is NOT PRO-tection. Never has been and never will be. The problem here seems to be that either the fire alarm sales engineer is misinformed or is more interested in selling their product than assuring the customer has what they need. Have you ever heard the old saying that when the only thing you sell is hammers, every problem looks like a nail?"

"Yes, I've heard that before," said the BYFPE, "But how does that apply here?"

"Like the hammer salesman seeing every problem as a nail, we run into the same thing in the fire protection industry. If you are a fire alarm guy, you think the answer to every problem is some type of fire detection and alarm system. If you sell sprinklers, the answer to every problem is a sprinkler system. If you sell clean agents, a clean agent extinguishing system solves every problem," explained the WOFPE.

"When there is only one tool in your toolbox, you can't effectively address the wide-ranging, site-specific needs of each facility. It's our job to ensure our client gets what they need to address the site-specific needs and conditions and nothing less. We don't want to burden them with systems and features that don't advance their fire protection objectives. But, we want to make sure their site-specific needs are met. And for that warehouse, a fire detection and alarm system won't do that."

"I gotcha," responded the BYFPE as he pushed himself out of the chair and started toward the door. "I'm going to go call the plant manager and talk to him about hammer salesmen."

#13 – Fire Alarm System Bobbles and Bangles

"You wanted to see me?" asked the Brash Young Fire Protection Engineer (BYFPE) as he entered the office of the Wise Old Fire Protection Engineer (WOFPE).

"Yes, yes, take a seat," said the WOFPE as he returned a book to the bookcase behind his desk. "I looked over the design brief you prepared for the new fire alarm system for International Amalgamated Industries and wanted you to walk me through a couple of things."

"Sure," beamed the BYFPE as he plopped down in the chair across from the WOFPE's desk. "I think I did a pretty good job incorporating all the design criteria discussed at the project kickoff meeting. What did you think?"

"Oh, you did a fine job on the design brief, and your writing has improved immensely in the past few months," said the WOFPE. "My questions concern some of the items listed as required features. I want to make sure we separate the 'need to have' features from the 'nice to have' features."

"What do you mean?" the BYFPE asked with a hint of suspicion.

"Where did the requirement for touch screen, computerized, graphic annunciators at three different locations come from?" questioned the WOFPE.

"The head of security, who is also the chief of the plant fire brigade, wanted them," explained the BYFPE. "He was at the last National Fire Protection Conference and said he saw them demonstrated and thought they were neat. He thought they would be perfect for quickly locating the area of an alarm and directing the fire brigade and public fire department to the fire. He likes the idea of touching a screen and having a floor plan pop up, giving you the exact location of the alarm. He's also got ideas to add other information on hazardous materials, people needing assistance, and other stuff to the screens. Is there a problem with that?"

The WOFPE thought for a moment, then responded, "No, there's nothing wrong if that is what they need, and an expensive interactive touch screen annunciator is the best way to accomplish it. But, unfortunately, people often respond to these 'gee-whiz' features because of how they look without considering the consequences."

"Consequences?" questioned the BYFPE as he draped his arms over the sides of the chair. "What consequences are there to using a state-of-the-art computer graphics annunciator?"

"Cost, for one thing," said the WOFPE. "That's a pretty important consideration."

"Well, the cost isn't a problem on this project," explained the BYFPE. "I talked to a couple of manufacturers, and both said that the equipment cost for all three computer graphics workstations wouldn't amount to one-half of one percent of the total project cost. The IAI people didn't even blink an eye at the cost."

"I'm not talking about just the equipment costs," replied the WOFPE. "I'm talking about the ongoing life cycle costs to keep it up-to-date. I agree that the cost of the equipment required would be relatively small based on the project's overall cost. But did you include the initial cost of paying someone to sit down and do all the 'golly-gee whiz-bang' programming to get everything to show up in the correct location on the floor plans? And, someone has to gather and enter all that other information you are talking about popping up on the floor plans."

"Well, we didn't discuss that," explained the BYFPE straightening himself in the chair.

"Add to that the cost of having someone sit down and change the programming every time they make a change in the facility," continued the WOFPE. "About how often are we over there to discuss changes they plan to make? From my estimation, that occurs at least once a month, if not more often. That could add up to a lot of money down the road."

"Yes, but they said they need this computer graphic annunciator because they need to be able to quickly pinpoint the location of an alarm and dispatch emergency responders to the location," explained the BYFPE.

Leaning back in his chair, the WOFPE pursed his lips and asked, "Is there a need for a computer-driven, touch screen, graphic annunciator, or is there a need to have the means to get first responders to the location of the alarm in the least amount of time?"

"Of course, the need is to get the first responders to the alarm location in the least amount of time," explained the BYFPE. He continued, "But how do you provide that capability without using a computer-based graphic annunciator?"

"You weren't yet part of the team, but a few years ago, we did a similar design for the Acme Ajax Corporation," explained the WOFPE. "Instead of slick looking, computer-based graphics for an annunciator, we used a notebook annunciator at each FACU and other locations in a plant covering over one million square feet."

"How is that different?" questioned the BYFPE. "You used a notebook computer, and we are using free-standing computer workstations."

"We didn't use a notebook computer. We used an actual physical notebook! You know those five-dollar, three-ring style binders sold in office supply stores," said the WOFPE.

"How in the heck can you use a notebook as an annunciator?" demanded the BYFPE

"Easy," explained the WOFPE. "We simply specified that the descriptive text label that shows up on the FACU had to end in a unique three-digit number. For example, activation of a smoke detector in Building 3, Room 217, on the second floor would show up as SD-B003-RM-217-295. The last three digits are the page number in a notebook we put together. There is a separate page for each initiating device. So you turn to page 295, and it shows a floor plan of the building with the location of the detector highlighted. We also added the other information you mentioned, such as types and amounts of hazardous materials and additional information useful for first responders. We made up several copies of the notebook. We put one at each security station, one at each FACU, and several at other locations for use by the public fire department and plant the fire brigade. If there are changes in the system or the information that needs to be 'annunciated,' it's a simple matter of updating the appropriate pages of the notebook. There's no programming involved. They use student interns to update pages as changes are needed. The last I checked, this system continues to meet the facility's needs, and there's no ongoing cost to maintain programming for computer graphics and touch screens."

The BYFPE mulled this over a moment as he moved toward the door. "This is part of your old school fire protection 'keep it simple stupid' rule, isn't it?" he queried.

The WOFPE smiled and returned to his work.

Chapter 5 – Manual Fire Suppression

#14 – Portable Fire Extinguishers: To Provide or Not to Provide

"Hey Boss," hailed the Brash Young Fire Protection Engineer (BYFPE) as he hurried down the corridor to catch up with the Wise Old Fire Protection Engineer (WOFPE). "Have a minute to talk about portable fire extinguishers?"

"Sure," said the WOFPE as he slowed his pace. "By coincidence, I was just heading to the lab to see a demonstration of a new extinguisher. Why don't you tag along? We'll discuss your questions and check out this new fire extinguisher together."

The two entered the fire lab, where technicians were preparing test fires in the demonstration room. Taking a seat at a table in the observation room, the WOFPE turned to the BYFPE and inquired, "OK, what's on your mind?"

"During my exit conference with the plant engineer at Amalgamated Industries yesterday, I was told someone in their safety department recommended the removal of all portable fire extinguishers since OSHA doesn't require them," explained the BYFPE. "He said the OSHA inspector told them if they keep the extinguishers, they have to maintain them even if they don't train anyone to use them."

"The plant engineer wants some ammunition to argue against removing all the extinguishers, but he knows facility management doesn't want to provide fire extinguisher training for everyone. So, I explained that the local fire code requires the provision of portable fire extinguishers. Still, it seems silly to provide them if they will never be used. Any suggestions on what else I can tell him?"

"Well, you hit the big one," said the WOFPE. "The fire department is not going to permit removal of the extinguishers in violation of the fire code. Keep in mind that OSHA is concerned only with the safety of the workers and has no concern with the building or continued business operations. That's why they don't care if you don't have extinguishers. They are concerned someone might be injured trying to fight the fire. So, they give the building owner the option of removing the extinguishers and directing the evacuation of all employees during a fire. As long as everyone gets out alive and unhurt, OSHA is happy, even if the building burns down and the company is out of business."

"That's nuts," exclaimed the BYFPE. "The fire code says the extinguishers have to be there and have to be maintained. But, why have them if no one will ever use them?"

"It's an interesting dilemma, isn't it," noted the WOFPE. "Provide and maintain extinguishers to comply with the fire code and the OSHA regulations. But instruct employees to ignore them because the company doesn't want to train anyone in their use. I agree, it sounds nuts, but it's a common scenario."

"OK, so what should we recommend?" questioned the BYFPE. "Keep them? Get rid of them? Train employees? Don't train employees? What's the answer?"

"We tell them what we always tell them," the WOFPE calmly explained. "Like any question about the necessary level of fire protection, we tell them they need to conduct a thorough, site-specific needs analysis to determine the best course of action. The needs analysis determines the fire response level and the options for meeting that need. Failure to conduct the needs analysis is an abdication of responsibility by facility management. Based on the hazards at Amalgamated, I would suggest that removing or ignoring the portable fire extinguishers is probably not the best course of action. They probably have twenty or twenty-five volunteer firefighters from area fire departments employed there. So, they already have a cadre of individuals with basic training using portable fire extinguishers. They could designate those individuals as the only employees to use the extinguishers while everyone else evacuates."

"Yearly extinguisher training and education would be relatively easy and inexpensive if it includes only those individuals. That would satisfy OSHA, the fire department, and the property insurance company. It would also give the facility an immediate trained response to a fire."

"Any management concerned with the fire safety of their employees and their facility would never rashly abandon the portable fire extinguishers without a thorough site-specific needs analysis. Tell them to check out Chapter 11, "Pre-Emergency Planning" in "**Industrial Fire Control Concepts**," for additional information."

#15 – Portable Fire Extinguisher Capabilities (Part 1)

"Hey Boss, thanks for sending me to that course at the fire academy this past weekend," gushed the Brash Young Fire Protection Engineer (BYFPE), sliding into a seat next to the Wise Old Fire Protection Engineer (WOFPE) in the lunchroom. "That was an experience everyone in fire protection should have."

Slowly stirring a packet of sweetener into a glass of iced tea, the WOFPE responded, "I'm glad you found the course useful. We've been sending people over there for years, and almost everyone seconds your assessment. So, tell me, what did you find most enlightening?"

"Easy, the hands-on fire extinguisher training," said the BYFPE. "You always think of fire extinguishers for use on a wastebasket fire or something burning in the kitchen, but we attacked and extinguished some big fires. I never realized how much fire you can put out with a portable fire extinguisher."

"I fully agree," said the WOFPE. "In college, I worked summers in an oil field as a roustabout. As part of their safety training, everyone had to undergo a day-long training on portable fire extinguishers. They set fire to excess crude oil in a pit and had us extinguish the fire with 20-lb dry-chemical extinguishers. I remember the first fire they set covered an area about 20 feet wide and 30 feet long. The heat was incredible, and I thought there was no way we would put it out using a fire extinguisher. But by the end of that day, not only could I put those fires out, I could do it while using only about one-half of the dry-chemical agent in the extinguisher. Like you, I thought it was a fantastic experience that paid off big time later in the summer."

"How did it pay off? Did you have a fire?" inquired the BYFPE.

"Well, a couple of weeks after the training, the volunteer fire department I was on got a mutual aid call for an 'oil well fire' one night. I was on one of about six engine companies called to the fire. It was the equivalent of a three-alarm response. Going down the road, we could see the glow of the fire in the sky from miles away. As we got to the scene, it was chaos. Fire trucks were everywhere. Firefighters were stretching hose lines. Portable tanks were being set up and filled by tankers. There were people everywhere."

"Leaking crude oil, ignited by the gas flame in the oil-water separator, spilled into the adjacent water pit and was burning fiercely. Oil field workers quickly shut down the supply of crude. So we were left with a water pit about forty feet long by twenty feet wide with burning crude oil floating on top of the water. It was a spectacular sight, particularly at night. Still, it wasn't much different than the training fires I had attacked a couple of weeks earlier."

"Our engine company captain, disgusted by the chaos and lack of organization, looked at me and pointed to a 20-lb dry-chemical extinguisher carried on the running board of the engine. 'Go ahead and put it out,' he said. I took the extinguisher off the engine and armed it. While everyone else was dragging around hoses and five-gallon buckets of foam, I walked up to the pit's edge, squeezed the handle, and swept the extinguisher nozzle across the surface of the fire, just like in training. Initially, it flared up, and then it went out. The whole exercise took less than ten seconds."

"It was weird. One minute there was this roaring hot fire with flames shooting forty feet in the air. The next minute the fire was out. Everything got quiet, and everybody just looked at me. The scene was lit only by the myriad of flashing lights on the fire trucks filling the road."

"The harried chief serving as incident commander looked at me and yelled, 'What do you think you're doing?' It was like he was mad that I put the fire out. Our captain was great. He acted like we did this type of thing all the time. He told the chief we were going back in service unless there was something else we could do for him."

"So, you put out a three-alarm fire with a fire extinguisher?" asked the BYFPE raising his eyebrows.

Picking up his sandwich, the WOFPE said, "Well, it wasn't actually a three-alarm fire. They sent that many fire companies because someone hit the panic button when they heard it was an 'oil well fire.' The lesson is that anyone with the appropriate training could have done the same thing. It's simply a matter of knowing the capabilities of fire extinguishers that comes only from actual fire training and experience."

#16 – Portable Fire Extinguisher Capabilities (Part 2)

"I've got a video that will interest you," said the Brash Young Fire Protection Engineer (BYFPE) to his mentor, the Wise Old Fire Protection Engineer (WOFPE). "It's of that fire at Acme Ajax we are supposed to look at tomorrow."

"Someone from the fire department told me an employee using a fire extinguisher saved the day," said the WOFPE.

"That's what the security video shows," explained the BYFPE. "One of the overhead doors on the loading dock is about halfway open. When an employee arrives with a 10-lb ABC dry-chemical fire extinguisher, the entire shipping area appears involved in fire. First, he discharges it through the side window and knocks the flames down. Then, when he sees a flare-up, he hits it again with an extinguisher off a maintenance truck. Other employees are arriving with extinguishers from other areas when the fire department gets there. The fire department extinguished and overhauled the fire before it got into the attached manufacturing area. It's amazing to see firsthand how much fire that 10-lb extinguisher knocked down."

"It just goes to show you how valuable a properly trained employee with a suitable fire extinguisher can be when a fire breaks out," observed the WOFPE. "Yet, no one wants to do fire extinguisher training, and every manager seems bent on getting rid of the fire extinguishers because their management spreadsheet says it saves money."

Chapter 6 – Automatic Sprinkler Systems

#17 – Sprinklers and Floor Coverings

"Hey, Boss," hailed the Brash Young Fire Protection Engineer (BYFPE) to the Wise Old Fire Protection Engineer (WOFPE). "I reviewed the current design analysis for the planned renovation of the old First Consolidated office building."

"What's the problem?" asked the WOFPE.

"The building code doesn't require sprinklers because it's an office building. But it seems like the perfect time to put sprinklers in the building, so I made that recommendation. Unfortunately, the new owners said sprinklers cost too much," said the BYFPE."

"How much is 'too much,' and to what are they comparing it?" asked the WOFPE.

"They didn't tell me the cost, but they said it was too expensive," explained the BYFPE. "I did check the cost estimating guides and looked at a couple of recent similar projects to determine an estimate based on a per square foot cost to install sprinklers. Using that information, I estimated it should cost no more than about $225,000. Any ideas on how to convince them that it's not 'too expensive' for a building of that size? Their project budget is about eight million dollars."

"Floor coverings, my young colleague. Floor coverings are the answer," explained the WOFPE.

"Have you been listening? What on earth do floor coverings have to do with the cost of installing sprinklers in the building?" questioned the exasperated BYFPE.

"They don't have anything to do with the cost of installing sprinklers," explained the WOFPE. "They offer a cost comparison point for anyone who says sprinklers are 'too expensive.' In my experience, the cost of installing sprinklers in an office building is a fraction of the cost of the floor coverings. Check the cost estimates and budget in the project design manual. I'd be willing to bet dollars to donuts they plan on spending two to four times as much on the carpet, tile, and other floor coverings than the total cost of sprinklers for the building. Think about it. After a fire, would you want to be the building owner trying to justify why they prioritized fancy floor coverings over installing a sprinkler system that protects the property, the people in it, and the continued business operations?"

#18 – Sprinklers and Water Damage

"Hey, Boss," said the Brash Young Fire Protection Engineer (BYFPE), "Is it true that there's more water damage in a building protected by sprinklers than in buildings without sprinklers?"

The Wise Old Fire Protection Engineer (WOFPE) looked up from the report he was reading, took stock of the BYFPE, and answered, "Of course it's true, that is if you are talking about percentages. So what prompts the question?"

The BYFPE explained, "Well, I just read an article in a builder's magazine, and they were talking about increasing requirements for installing sprinklers in apartments and condos. The article argued that sprinklers increase insurance costs because statistics show that the water damage after a fire in a sprinklered building is usually about ten times the actual fire damage."

"That's probably true as well," said the WOFPE. "But what they are not giving you is the total damage with and without the presence of sprinklers. The major industrial property insurance carriers report that the total damage, meaning fire and water damage combined, is eight to ten times greater for incidents occurring in buildings without automatic sprinklers than in buildings protected by sprinklers. Think about it. Of all the fires you've investigated, which ones had the greatest total damage? Those with sprinklers or those without sprinklers?"

"Those without sprinklers, of course," replied the BYFPE. "Most of them burned the building down."

The WOFPE elucidated, "Exactly, a lot of people, including builders and developers, have used this specious argument against sprinklers for years. Do you think the insurance companies, who pay for the fire **AND** water damage, would continue requiring sprinklers if it costs them more?"

"Probably not," responded the BYFPE.

"I can guarantee you not," expounded the WOFPE. "Water damage means there was something left to damage with water. Remember, the only fire in which there is no water damage is a total burnout. If the fire destroys everything, there is nothing left to be damaged by the water, so it's all recorded as damage from the fire."

#19 – Hydrophobia

"Hey, Boss," said the Brash Young Fire Protection Engineer (BYFPE) as the Wise Old Fire Protection Engineer (WOFPE) entered the sprinkler lab. "The plans for that new IT center at Conglomerated International Industries' global headquarters show they are planning a double interlocked, combination dry-pipe, preaction sprinkler system with cross-zoned heat and smoke detection to activate it. I know that's not what you recommended, but the IT manager contends it's necessary to protect her critical IT equipment."

"Sounds as though someone over there is suffering from an acute case of hydrophobia," said the WOFPE as he examined a dry-pipe valve in the wet lab.

"Hydrophobia?" asked the BYFPE with a quizzical expression.

"Yes, it's an irrational fear of water," explained the WOFPE. He continued, "IT managers often suffer from hydrophobia when someone talks about installing sprinklers in their computer center. They believe in the Hollywood depiction of sprinkler systems. They think if there is a puff of smoke, the sprinklers will wash their expensive equipment into the parking lot. So instead of protecting their precious equipment, they end up doing everything possible to keep from protecting it."

"How can you say that?" countered the BYFPE. "They are probably spending ten times more on that double interlocked, combination dry-pipe preaction system than it would cost for a wet-pipe system. So, she must be concerned about protecting the equipment."

The WOFPE manipulated a valve to start the setup of a dry-pipe valve for a demonstration. "Think about it. Instead of a wet-pipe system that simply requires a fire to fuse a sprinkler before it starts suppressing the fire, their proposed system sets up multiple hurdles and hoops you must traverse before any water gets on the fire. Of course, the fire keeps getting bigger while you jump those hurdles and go through those hoops. In their system, they first have to activate a smoke detector. Then it requires activation of a heat detector. That opens the preaction valve, but you still don't get any water on the fire. Next, the fire must fuse a sprinkler, releasing the air pressure before the dry-pipe valve trips. Once the dry-pipe valve trips, water has to fill the piping and expel the air, and finally, after who knows how long, you get water discharging from the system. By that time, whatever is on fire, and whatever is surrounding it is very seriously damaged or even destroyed."

"On top of that, they must inspect, test, and maintain a dry-pipe valve, a preaction valve, a heat detection system, a smoke detection system, and all the control equipment needed to meld those systems together. You've increased the component count by a factor of ten, so you've increased the potential for failure by about the same magnitude. Does that sound like they are really concerned about protecting their precious IT equipment?"

"I think her real concern is a sprinkler being knocked off, or popping off, and wetting down the equipment," posited the BYFPE. "She thinks the combination dry-pipe preaction system minimizes the potential for accidental water damage."

"And, of course, you can dispel that fear by explaining that she probably has a better chance of being struck by lightning while she's on the way to pick up her lottery jackpot winnings than a sprinkler just 'popping off' for no reason," countered the WOFPE. "Hopefully, one day, we will have a cure for the affliction of fearing the water used to extinguish a fire more than the fire itself."

#20 – Existing Pipe Schedule Sprinkler Systems

"Step in here a moment," said the Wise Old Fire Protection Engineer (WOFPE) to his colleague, the Brash Young Fire Protection Engineer (BYFPE), as he motioned him into a chair in the office. "I need you to head over to the Acme-Ajax Machinery plant. They are considering leasing an existing warehouse building across the street for extra manufacturing space."

"Sure, Boss," said the BYFPE. "What's the specific issue we are examining?"

"The building department told them that because of the change in occupancy, the building must comply with current code requirements," explained the WOFPE. "The fire department said it would require a new sprinkler system since the existing system is a pipe schedule system installed in the late 1970s. They said it had to be a hydraulically designed system. We were involved in an internal examination of the sprinklers about three years ago when the current owners purchased the building. At that time, the interior condition of the piping was fine, so there shouldn't be any concerns with obstructions."

"So are we going to design new sprinkler systems?" inquired the BYFPE.

"Not exactly. You need to prepare sketches of the sprinkler systems showing pipe sizes and lengths so we can run some hydraulic calculations to see what density the existing systems can produce," explained the WOFPE. "With the strong water supply in that area, I'll bet we can get the density needed from the existing pipe schedule system. The last water flow test we did in that area showed we had the equivalent of the Ohio River under the street at 150 psi. I know there are other buildings in that area where we've calculated the existing pipe schedule systems and were able to get a density of 0.30 GPM/sq.ft. over 3000 sq.ft. without changing any piping in the building. If we can do that here, it will save our client a lot of time and money, provide the protection our client needs, and satisfy both the building and fire departments."

"But, Boss," questioned the BYFPE. "Will the fire department accept that as equivalent to a hydraulically designed system?"

"Of course they will," explained the WOFPE, "Keep in mind that hydraulic calculations are a means of determining the pipe sizes needed in a new system to produce the desired density AND a means of analyzing an existing piping system to determine its capabilities based on the available water supply. Whether the calculations are to design a new system or evaluate the capabilities of an existing system, the layout and operation of the system are the same. Hydraulic calculations are simply one of the tools in our fire protection toolbox that we can use to assure the sprinklers are capable of meeting the challenge posed by the hazards they protect."

#21 – A Note About the Precision of Hydraulic Calculations

"Hey, Boss," said the Brash Young Fire Protection Engineer (BYFPE) to his mentor, the Wise Old Fire Protection Engineer (WOFPE). "The insurance company rejected those calculations I prepared by hand for the existing sprinklers in the Acme Ajax Warehouse. You know the one where we thought the existing sprinklers would be adequate based on the strong city water supply there."

"Why did they reject them?" inquired the WOFPE.

"They said that when they ran the system through their hydraulic calculation program, they came up with a difference in the demand of 13 GPM and 1.1 psi from what I calculated," explained the BYFPE. "My hand calculations showed we needed 734 GPM at 64 PSI. They said their program calculated a total demand of 747.278 GPM at a base of riser pressure of 65.121 psi. Still, they wouldn't provide me with a copy of their calculations to compare to mine because they said they didn't want the liability associated with 'designing' a sprinkler system." The BYFPE made air quotes around the word "designing."

The WOFPE punched the keys on the calculator on his smartphone and turned to the BYFPE. "So, your hand calculations varied less than two percent from their computer calculations, but they rejected them?"

"Afraid so," said the BYFPE. "The thing is, no matter which calculations you might use, the water supply is strong enough to provide a factor of safety of more than 150 percent. I didn't think the calculation methods were so precise that such a small variance made any difference."

"It doesn't," explained the WOFPE. "Except to people who don't understand that the purpose of the calculations is to confirm that water supply can support the necessary sprinkler density. Some people get so wrapped up in the numbers they lose sight of the fact that the basis of the entire hydraulic calculation method is a set of assumptions we make that simplify the physics of the flow in the system so we can mathematically calculate an estimate. Remember your discussions in engineering school about significant digits and levels of precision?"

"Sure I do," said the BYFPE.

"Just think about it," continued the WOFPE. We can do a water supply flow test using the latest equipment calibrated in the laboratory immediately before the test. Because we know what we are doing, our test results are probably accurate to within thirty to fifty gallons per minute with a residual pressure accurate to within one or two psi. So now we give that information with a margin of error in the flow somewhere between thirty and fifty gallons per minute and a margin of error in the residual pressure reading of one or two pounds per square inch to someone with a computer-based hydraulic analysis program. And miracle of miracles, they develop hydraulic calculations to the fourth decimal place. Does that sound reasonable to you?"

'No, it doesn't," responded the BYFPE.

"It's like I said," explained the WOFPE. "Some people get so caught up in the numbers and the workings of the computer program that they lose sight of the fact that what they are calculating is an ESTIMATE of the required pressure and flow under one specific set of circumstances. If you were careful, your manual calculations and the computer-based calculations prepared by the insurance company probably have about the same level of precision. The demand for that system is not 734 GPM at 64 psi as you calculated, and it's not 747.278 GPM at 65.121 psi as they calculated. In reality, it's probably about 730 to 750 GPM at a pressure of about 62 to 67 psi if a fire occurs in the calculated area and operates only the sprinklers included in the calculations."

Hydraulic Analysis

Job Number: 2539 - Acme Ajax Offices
Report Description: Ordinary Group I (System #5)

Pipe Type Downstream Upstream	Diameter Elevation	Flow Discharge	Velocity K-Factor	HWC Pt	Pn	Friction Loss Fittings	Length Eq. Length Total Length	Pressure Summary	
····· Route 1 ·····									
DR 3200 3033	1.0490 25'-4 26'-10	21.17 21.17	7.86 8	120 7.000 8.114		0.144514 Sprinkler, 2E(2'-0), PO(5'-0)	3'-2 9'-0 12'-2	Pf Pe Pv	1.756 -0.641
BL 3033 3034	1.6100 26'-10 26'-10	2.37	0.37	120 8.114 8.117		0.000313	8'-0 8'-0	Pf Pe Pv	0.003
BL 3034 3035	1.6100 26'-10 26'-10	23.54 21.17	3.71	120 8.117 8.292		0.021846 Flow (q) from Route 2	8'-0 8'-0	Pf Pe Pv	0.175
BL 3035 3036	1.6100 26'-10 26'-0	44.93 21.38	7.08	120 8.292 10.336		0.072199 Flow (q) from Route 4 3E(4'-0)	11'-3½ 12'-0 23'-3½	Pf Pe Pv	1.683 0.361
BL 3036 3017	1.6100 26'-0 26'-0	68.46 23.54	10.79	120 10.336 12.703		0.157397 Flow (q) from Route 13 PO(8'-0)	7'-0½ 8'-0 15'-0½	Pf Pe Pv	2.367
CM 3017 3018	3.0680 26'-0 26'-0	94.29 25.83	4.09	120 12.703 12.858		0.012315 Flow (q) from Route 14	12'-7 12'-7	Pf Pe Pv	0.155
CM 3018 3019	3.0680 26'-0 26'-0	147.90 53.61	6.42	120 12.858 13.161		0.028320 Flow (q) from Route 5	10'-8½ 10'-8½	Pf Pe Pv	0.303
CM 3019 3020	3.0680 26'-0 26'-0	187.03 39.13	8.12	120 13.161 14.999		0.043720 Flow (q) from Route 9 4fE(5'-0)	22'-0½ 20'-0 42'-0½	Pf Pe Pv	1.838
CM 3020 3008	3.0680 26'-0 26'-0	220.24 33.21	9.56	120 14.999 17.115		0.059160 Flow (q) from Route 15 2fE(5'-0)	25'-9½ 10'-0 35'-9½	Pf Pe Pv	2.117
CM 3008 3009	3.0680 26'-0 26'-0	247.64 27.40	10.75	120 17.115 17.771		0.073489 Flow (q) from Route 3	8'-11 8'-11	Pf Pe Pv	0.656
CM	3.0680	277.80	12.06	120		[illegible]	15'-9½	Pf	[illegible]

#22 – Establishing Sprinkler Design Criteria

"Hey, Boss," said the Brash Young Fire Protection Engineer (BYFPE) as he poked his head into the office of the Wise Old Fire Protection Engineer (WOFPE). "I was looking over the plans for the new Amalgamated Motors facility, and they've specified a sprinkler design of 0.45 GPM/sq.ft. over a 5000 sq.ft. operating area for all manufacturing areas, and a density of 0.60 GPM/sq.ft. over all storage areas along the production line. That's like double what the codes require. What gives?"

The WOFPE motioned the BYFPE in, pointed toward a chair, removed his glasses, and rubbed his eyes. Then, he began, "Do you remember the case studies of past fires you reviewed as part of your last training lab class?"

"Sure I do," said the BYFPE. "They were one of the best things about the class."

"The case study of the distribution center that burned to the ground was an Amalgamated Motors facility," explained the WOFPE. "That fire crippled Amalgamated Motors for months, and they looked at it and said, 'Never again will we allow that to happen.' So, they completely overhauled their corporate fire protection program to address the problems uncovered after the fire. Do you remember what happened in that fire?"

The BYFPE thought for a moment and then said, "If I remember correctly, the facility and the sprinklers were originally designed for mostly metal parts in cartons, but as technology changed over time, they ended up with a lot of plastic and other combustible parts in the facility. The original sprinklers weren't designed for those hazards, so the facility burned to the ground. There were some other factors, but that's the gist of the story."

"Exactly," illuminated the WOFPE. "After the fire, they recognized that 'hazard creep' had invaded most of their facilities, and they had hazards that the existing sprinklers could not protect. So to avoid future problems, they instituted a policy of using a density of 0.45 GPM/sq.ft. over a 5000 sq.ft. operating area for all manufacturing areas regardless of what the codes permitted, and a density of 0.60 GPM/sq.ft. over all storage areas. They wanted to ensure they were never again the victims of hazard creep."

The BYFPE questioned, "I like the idea of such a large safety factor, but doesn't that cost them a lot of money?"

"Not as much as a disaster like the distribution center fire," explained the WOFPE. "It does cost more, but the real cost of a sprinkler system is the water supply and the labor to hang the piping. A two-inch pipe does cost more than a one-inch pipe, but not that much more. The real cost is the labor to install it, and there isn't that much difference between the cost of labor to hang a one-inch pipe and a two-inch pipe. The water supply from the multiple fire pumps is required regardless. So they figure they might as well take advantage of the water supply and provide the highest density practical even if they don't need it right now. Other organizations could benefit a great deal from this approach."

Chapter 7 – Fire Protection Water Supplies

#23 – Water Supply Analysis Methods

"I've got an assignment for you," said the Wise Old Fire Protection Engineer (WOFPE) as he motioned the Brash Young Fire Protection Engineer (BYFPE) into his office.

"Sure," said the BYFPE. "So, where am I going?"

"Tomorrow morning, you are meeting with the plant engineer at Hi-Tek Electronics to discuss their fire protection water supply," said the WOFPE. "She's upset about the cost of their new fire pump and 300,000-gallon reservoir. She said when the fire department came through the facility, they said a fire pump wasn't necessary. She just wants to know they aren't wasting money."

"We provided them with a complete analysis explaining why they need the pump and reservoir," protested the BYFPE. "I agree that they don't need a pump to comply with the fire code, but they need one to address the facility's site-specific risk management goals and enabling objectives. I thought we explained all of this?"

"We did, but I guess with a half-million dollars on the line, she wants to make sure it's the correct decision," said the WOFPE. "I think we just need to remind her that the methods used by the public fire officials to determine the required water supply are not based on protecting their facility but rather on protecting the public from their facility. The fire code is only concerned about ensuring that a fire does not spread from their property and affect any other properties. It also doesn't care if they are still in business after the fire."

"Yes, but didn't their property insurance carrier also recommend a pump?" questioned the BYFPE.

"Industrial Mutual did recommend the installation of an electric booster pump, but not the reservoir," responded the WOFPE. "Discussing the booster pump with the corporate fire protection engineer brought out the fact that the public water supply to the facility has been impaired three times in the last two years because of main breaks, leaving the facility without a water supply for a significantly long time. Hi-

Tek considers this site critical to its supply chain. A problem there affects multiple downline facilities. The corporate risk management department decided it would be prudent to install a pump and reservoir, so the facility is never without water, even during a public water supply outage."

"Determining water supply requirements isn't always a simple matter," said the BYFPE. "I can see why a lot of people get confused."

#24 – Gate Valves vs. Butterfly Valves

"Hey, Boss," questioned the Brash Young Fire Protection Engineer (BYFPE) as he sat down in the office of the Wise Old Fire Protection Engineer (WOFPE). "I was asked a question today that I'm unsure how to answer."

"Let's hear it," said the WOFPE removing his glasses and setting them on the sprinkler drawings he was reviewing.

"A contractor cornered me and asked me to explain why I recommended they install gate valves instead of butterfly valves for sprinkler control," explained the BYFPE. "He said as long as it's a valve listed for fire service use and it meets NFPA 13, it should be permitted. He said butterfly valves are smaller, take up less space, and cost less than a gate valve, like an OS&Y valve."

"And you told him what," asked the WOFPE.

"I told him I had to get on a conference call but would call him to talk later," replied the BYFPE sheepishly.

"OK, let's look at the differences and why we prefer a gate valve over a butterfly valve," said the WOFPE. "Both have advantages and disadvantages. The butterfly valve is smaller, it does cost less, it takes up less space, and there are butterfly valves listed for fire service use. On the other hand, the OS&Y valve is larger, requires more space, and costs more, so why would we prefer it?"

"Probably because of the friction loss. A gate valve, like an OS&Y valve, has less pressure loss through it," explained the BYFPE.

"True," agreed the WOFPE, "but let's look at this diagram." He pulled out an illustration of each type of valve and laid it on the desk. "One of the big advantages of the gate valve on the left is that if something, like a small rock or debris, gets caught in the bottom of the valve seat, we can probably still get the valve closed enough to stop the flow of water and be able to work downstream of the valve. This eliminates the need to shut down a larger portion of the water supply system to isolate a problem."

"If that same piece of debris lodges in the butterfly valve seat on the right, it holds the disc open around the entire diameter. The volume of water it allows through makes it impossible to work down steam without closing more valves and shutting down more of the water supply, with a greater impairment of the fire protection water supply."

"Ah," said the BYFPE, "I can now give the contractor TWO reasons why we prefer the gate valve over the butterfly valve. First, the gate valve creates less pressure loss. And secondly, the use of a gate valve provides some level of risk reduction during an impairment."

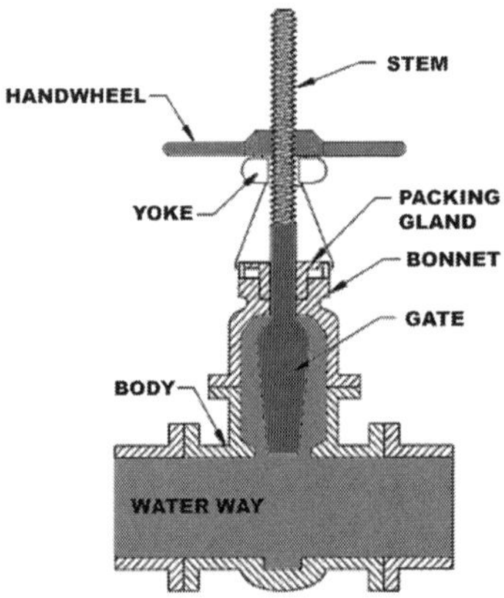

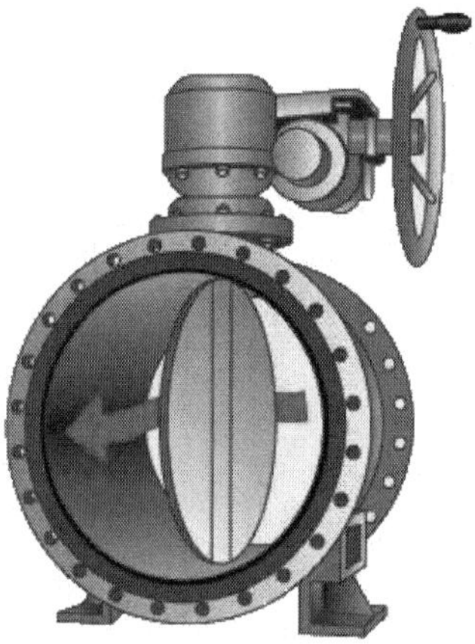

#25 – Backflow Prevention Assemblies

"I saw your report on the Amalgamated Industrial site over on the west side," said the Wise Old Fire Protection Engineer (WOFPE) to the Brash Young Fire Protection Engineer (BYFPE). "Let's talk about it for a few minutes."

"OK, what's up?" said the BYFPE.

"What worries me is this comment that says, since the last survey of the facility, they installed backflow prevention assemblies on each sprinkler system," explained the WOFPE. "Yet, I don't recall seeing any plans or calculations for the work."

Quizzically the BYFPE inquired, "Plans? Calculations? Just for the installation of a backflow preventer?"

The WOFPE inquired, "How is the water supply over there? Isn't it considered marginal because of the relatively low pressure?"

"We're on the borderline for a few of the sprinkler systems. In some cases, we had to upsize some of the piping to get to that point," explained the BYFPE. "But, what does that have to do with the backflow preventers?"

"Here's a catalog sheet for a typical backflow prevention assembly listed for fire protection service," said the WOFPE holding the sheet out to the BYFPE. "What does it show as the friction loss through an 8-in reduced pressure backflow prevention assembly at a flow of 850 GPM? Isn't that about the average demand for the systems at Amalgamated?"

The BYFPE carefully studied the manufacturer's specification sheet. "It looks like the loss would be about . . . eight or nine PSI at a flow between 800 GPM and 900 GPM." He studied the sheet more intently as the problem dawned on him. "Oh, I see what you mean. We don't have an extra eight or nine PSI over there. A loss of two or three PSI means we are below the pressure required for the sprinklers, and they are already marginal."

"Exactly," exclaimed the WOFPE. "Any change, even something that seems as innocuous as installing a backflow prevention assembly, can have a major impact on the ability of the fire protection systems to control a fire. Therefore, all proposed changes must be carefully reviewed and analyzed to determine the effect on the facility's overall fire control system."

The BYFPE excused himself, saying, "I need to make a phone call and revise a report."

#26 – Fire Service Connection Standby Fees

"Hey Boss," hailed the Brash Young Fire Protection Engineer (BYFPE) to the Wise Old Fire Protection Engineer (WOFPE). "The plant manager for Acme Ajax says they don't plan to install sprinklers in that long-term leased warehouse they moved into at the industrial park. He says that saves them $32,500 the first year and $12,500 each succeeding year."

"Enlighten me," said the WOFPE while motioning the BYFPE to give him the story.

The BYFPE continued, "He said the water company told him if they install sprinklers, they have to pay a $20,000 tap fee for a new 10-inch connection to the city main to feed the sprinklers. He'd then be billed $12,500 yearly as a 'standby' fee to maintain the sprinkler feed. Over ten years, that's almost $150,000 in fees just for the sprinkler system. If he doesn't install sprinklers, he doesn't pay anything."

"Ironic isn't it," observed the WOFPE. "If you take fire protection seriously and don't abdicate responsibility for protecting your facility, the public water authority kicks you in the face. But, on the other hand, if you don't install sprinklers and simply rely on the public fire department to 'protect' your facility, you don't pay anything."

"Why is that?" questioned the BYFPE.

"Total ignorance, my young colleague, total ignorance." replied the WOFPE. "Think about it. If there are no sprinklers and the warehouse catches fire, they call the fire department. The fire department shows up to find the building totally involved in fire. It becomes a surround and drown operation to keep the fire from spreading to surrounding buildings or properties. Ultimately, they tie up multiple fire companies for several hours and pump hundreds of thousands of gallons of water onto the fire. And they don't pay a dime for the water. If they install sprinklers designed for the hazards in the building and there is a fire, it may take only a few hundred gallons of water to control the fire. Still, they are charged $20,000 just for the privilege of connecting to the public water supply to protect their property and another $12,500 per year to maintain that connection."

"That's crazy," cried the BYFPE. "They've got it all bass-ackwards."

Sighing, the WOFPE replied, "That's what happens when bureaucrats make technical decisions. The owner that installs sprinklers reduces the demand on the public water system and the public fire department and ends up paying for all the owners who don't do anything to protect their property."

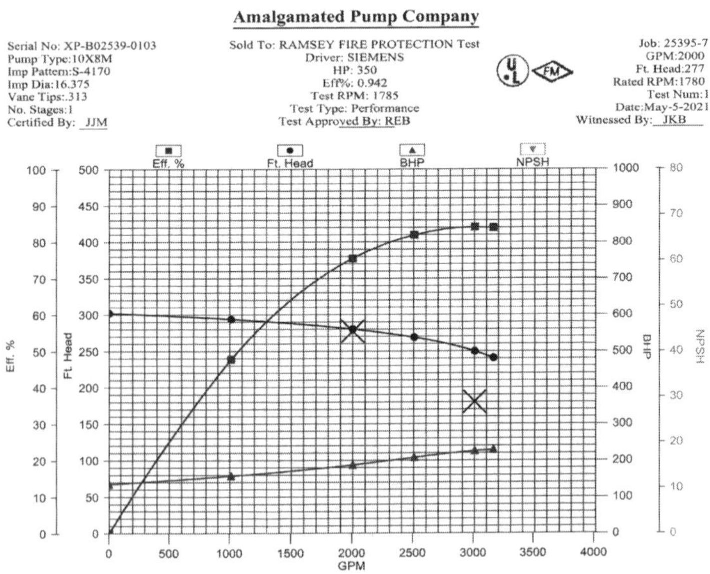

Chapter 8 – Fire Pumps

#27 – Fire Pump Certified Curves and Nameplate Data

"Come and sit for a minute," said the Wise Old Fire Protection Engineer, as he motioned the Brash Young Fire Protection Engineer (BYFPE) into a chair at the small conference table in his office. "Let's talk about the fire pump test you conducted last week at Hi-Teck International."

"OK," said the BYFPE warily. "Did I do something wrong?"

"I just want to ensure the pump performs as required," said the WOFPE. "But your plot of the test results just shows the minimum performance criteria from NFPA 20. We should be plotting the test results against the certified curve provided by the pump manufacturer."

"I know," answered the BYFPE. "But they didn't have a copy of the original curve, so I just plotted against the NFPA minimums."

"That's not unusual because most building owners don't recognize the need to maintain the pump documentation for the life of the pump," explained the WOFPE. "Fortunately, if you can get the pump's serial number off the nameplate, you can call or email the manufacturer. They usually retain the documentation on every pump they manufacture and are happy to provide a copy of the certified curve. Then you can plot the performance test against the original certified performance curve."

#28 – Tending a Running Fire Pump

"Hey, Boss," hailed the Brash Young Fire Protection Engineer (BYFPE), observing the Wise Old Fire Protection Engineer (WOFPE) walking toward his car in the parking lot. "Do you have time for a quick question before heading home?"

"Sure," said the WOFPE as he tossed his bag in the car. "What's on your mind?"

"First thing in the morning, I've got the acceptance test for a new diesel fire pump at International Amalgamated," explained the BYFPE. "I was preparing for the test and noticed that you changed my spec to say the automatic shutoff feature on the fire pump controller is to be disabled. What gives? As far as I know, nearly every UL-listed fire pump controller includes that feature. So why not use it?"

"It's simple," said the WOFPE. "We never want a fire pump running unattended. The Security Chief at Amalgamated has spent a lot of time pounding into her people that when they receive a pump running signal, they must immediately and without any delay get somebody to the pump house to attend the pump. If it shuts down automatically after running for thirty minutes, no one is in a real hurry to get to the pump house."

"If the pump is going to shut down automatically, what's the hurry to get there?" quizzed the BYFPE. "Why would a listed controller even have the feature then."

The WOFPE crossed his arms and leaned against his car. "I don't know. Convenience? A sales gimmick? Customer request? I'm not sure. But, we always recommend arranging all fire pumps for manual shutdown only. We want the pump running to be an inconvenience. We want to force someone to get to the pump house to attend the pump while it runs."

"OK, but what do I tell the plant engineer? He wants automatic shutdown enabled and says it's his pump, not security's," questioned the BYFPE.

"Ask him if he understands that his fire pump is the only piece of equipment in the facility designed to run to self-destruction," suggested the WOFPE. He continued, "Tell him about the incident at

Acme Ajax a couple of years ago where the diesel fire pump started and ran unattended because everyone thought it would shut down by itself. The only problem was that the oil pump failed about ten minutes after it started. They finally sent someone to the pump house when they got a low oil pressure signal. When they arrived, the pump house was full of smoke, and pistons were landing in the parking lot. The engine blew up. Lubricating oil and diesel fuel spewed through the pump house and was burning. The fire destroyed the engine and controller and severely damaged the pump house. They were without a water supply for weeks while repairs got the pump back in service. They ended up paying to have a fire department pumper stand by at their facility until the new fire pump was in-service. Have him call the Acme Ajax plant engineer and get his views on automatic shutdown. Then see if he still wants the automatic shutdown feature enabled."

#29 – Electric Fire Pump Power Supply Reliability

"Hey, Boss?" inquired the Brash Young Fire Protection Engineer (BYFPE) of her mentor, the Wise Old Fire Protection Engineer (WOFPE). "What's this note you put on my preliminary specifications for the new fire pumps for the Amalgamated Chemical warehouse facility?"

"Yes, let's talk about that spec for a minute," said the WOFPE motioning for the BYFPE to take a seat. "You're specifying three electric motor-driven pumps for a facility being built in Texas, correct?"

"That's right," confirmed the BYFPE.

"Did you investigate the reliability of the electrical power supply to the site?" questioned the WOFPE.

"Of course, I did," responded the BYFPE somewhat indignantly. "The website for the power company in that area touts that they have a record of power availability over the past ten years that exceeds ninety-nine percent. That sounds pretty reliable to me."

"OK," said the WOFPE. "There are 24 hours in a day and 365 days in a year, so over ten years, that's 87,600 hours. So, if they provided power for ninety-nine percent of that time, that still leaves 876 hours without power, which equates to 36.5 days. Did they happen to say how many successive days some of their customers were without power during the Texas deep freeze in the Winter of 2020?"

"No, not specifically," said the BYFPE.

"Well, we have some clients in that area that were without power for almost two weeks during that incident," explained the WOFPE. "That's a long time to be without fire protection. And that is just for one incident. That area has experienced other widespread, long-term outages due to hurricanes."

"I guess I didn't realize the Texas power grid was so vulnerable," admitted the BYFPE.

Taking a deep breath, the WOFPE continued, "It's not just Texas. In recent years, there have been hurricanes, ice storms, blizzards, floods, wildland fires, and other natural disasters that have left vast areas of the country without power for much more than the maximum four, or eight, or ten hours in a year specified by NFPA 20 and various AHJs. In recent times large portions of not only Texas but Oklahoma, California, the East Coast, and the Gulf Coast have experienced widescale power outages that disqualify them as areas with reliable electric power as defined by NFPA 20. It's something that needs careful consideration when evaluating the power supply for an electric fire pump."

"So, you think I should specify diesel-driven pumps instead of electric?" probed the BYFPE.

"I think you need to more carefully evaluate the anticipated reliability of the electrical supply and base your decision on the needs of the facility," explained the WOFPE. "Maybe you can use electric, maybe you should use diesel, and maybe you should consider having some of each. It all depends on the importance of maintaining the protection and the site-specific needs and conditions of the facility."

#30 – Electric Versus Diesel Fire Pumps

"Hey, Boss, when a fire pump is needed, why is it that you always seem to recommend a diesel engine-driven pump instead of a more economical electric pump?" inquired the Brash Young Fire Protection Engineer (BYFPE) of the Wise Old Fire Protection Engineer (WOFPE). "Doesn't looking out for the best interests of our clients extend to making sure they don't waste their money?"

"Wow!" exclaimed the WOFPE. "There's a lot in that question that I think we need to flesh out. First, acting in the best interests of our clients DOES mean we must consider the economic implications of our recommendations. Still, it doesn't mean we automatically recommend the least expensive option. Least cost does not necessarily equate with the best value. It is one factor to consider. But, it's our job to present the options that best meet the facility's fire protection, risk management, and site-specific needs so the client can make an informed decision."

"OK, I understand it's not as simple as I may have tried to make it," said the BYFPE. "But let's take the water supply problems at the Acme Ajax plant. How do I justify recommending a new diesel engine-driven fire pump that costs two or three times what an electric motor-driven fire pump of the same rating costs?"

The WOFPE sighed and slowly let out his breath. "Let's take it one step at a time. First, why are they installing the fire pump?"

"Because the pressure available from the public water supply is grossly deficient and can't meet the sprinkler demand in the new warehouse," said the BYFPE with a bit of exasperation as if that should be obvious to anyone familiar with the facility.

Trying to temper his frustration, the WOFPE slowly and deliberately asked the BYFPE, "And, how important is the warehouse to Acme Ajax?"

"It's critical," explained the BYFPE. "Right now, it's the only place they have to store their intermediate feed material for several plants. That's why they realize the fire pump is needed."

"So, it's a critical facility, the loss of which would put a real hurt on their production," said the WOFPE. "And, you plan to provide an electric motor-driven fire pump because you say it's cheaper? Does that include the cost of providing a dedicated electrical supply for the pump separate from the warehouse? And, what about electrical supply reliability? Twice in the past three years, we've had major storms that knocked the power out for several days in that area."

"Well, they could put in a diesel emergency generator as a backup power supply for the pump," suggested the BYFPE.

"Let me get this straight," said the WOFPE. "You want to provide an electric motor-driven fire pump because it's less expensive than a diesel, but then install a diesel engine-driven generator to provide emergency power for the electric fire pump?"

Comprehension dawned on the BYFPE as he said, "That does sound absurd when you say it out loud, doesn't it?"

"I prefer diesel engine-driven fire pumps for many, but not all applications," explained the WOFPE. "The site-specific fire protection and risk management goals drive the decision as much as the economics. However, where the situation demands the highest degree of reliability and availability, a self-contained diesel engine-driven fire pump installation, not dependent on the facility infrastructure or the public electric utility, best meets the need."

#31 – Fire Pump System Complexities

"Do you have that design proposal for renovation of the maintenance hangar at Universal International Airport ready for me to review?" asked the Wise Old Fire Protection Engineer (WOFPE) of the Brash Young Fire Protection Engineer. "We didn't win the job to prepare the Request for Proposals (RFP) because they said we were too expensive, but maybe we'll be more successful in getting the design contract."

"I submitted a couple of RFIs (requests for information) to the contracting officer for which I've not yet received answers," explained the BYFPE. "And the fire protection system design for the entire hangar bay hinges on them."

"Really," said the WOFPE. "What's the problem?"

The BYFPE explained, "The RFP says they expect that the new foam system for the hangar bay requires higher pressures than available from the existing fire pumps. So instead of replacing the pumps, it says to replace the impellers with larger impellers producing more pressure."

"You're kidding, right?" questioned the WOFPE. "You can't just pull one impeller out and stick in a larger one. Did they happen to provide the analysis on which they based that decision?"

"It wasn't part of the RFP," responded the BYFPE. "One of the RFIs I prepared asks for a copy of the analysis. At the pre-proposal walkthrough, I took a photo of the pump nameplates showing the serial numbers and called the manufacturer."

"And the manufacturer said what?" questioned the WOFPE.

"I talked to the engineer involved with the original installation, so he was quite familiar with the facility," explained the BYFPE. "After he stopped laughing, he said not only won't a larger impeller fit in those pumps. He said it would take more horsepower to drive each pump even if larger impellers fit, so we need to replace all the drivers. Then, increasing the horsepower of the drivers requires new controllers rated for the higher horsepower drivers. The bottom line is that they will have to replace all three pumps, drivers, and controllers. I don't think that is something the airport authority was planning. That's what my RFI addresses. I thought maybe I was missing something."

The WOFPE responded, "You didn't miss anything. The engineer of record didn't do their job. A fire pump installation is a system of components carefully selected and engineered to provide a specific level of performance. Any competent designer should know that you can't alter one part of the system without evaluating the cascading consequences of the proposed change. This will be a budget-buster for the project. I wonder if the airport authority still thinks accepting the low bid proposal to develop the RFP was in their best interest?"

#32 – Fire Pump Starting Pressure

"You look perplexed," said the Wise Old Fire Protection Engineer (WOFPE) to the Brash Young Fire Protection Engineer (BYFPE). "What are you working on?"

The BYFPE contemplated an answer for a second and then said, "I'm trying to figure out the starting and stopping pressures for a new fire pump installation at that old industrial mill east of town. They have some 70- and 80-year-old fire mains, and the plant engineer fears the fire pump will blow them up."

"So, you want the pumps set to start at the highest pressure possible," offered the WOFPE. "That way, there's less stress on the old mains."

"How do you figure that?" questioned the BYFPE. "Logic says setting the start pressure as low as possible would be less stress on the piping because the pumps wouldn't run as often."

"But when they start, the underground piping is hit with a pressure surge or water hammer that is more likely to cause a break," explained the WOFPE. "Let's say the fire pump churn pressure is 150 PSI. If the fire pump start pressure is set for 60 PSI, it might not run as often, but it hits the underground piping with a 90 PSI pressure spike when it starts. If there are sections of weak piping, it's not the 150 PSI churn pressure that causes a break. It's that sudden pressure surge that does it."

"So, what's the solution?" asked the puzzled BYFPE.

The WOFPE continued, "Set the pump to start at or above the churn pressure. That minimizes any water hammer or pressure surge when it starts. For example, if the pump churn pressure is 150 PSI, set it to start at or above 150 PSI. If you set the jockey pump to start at 165 PSI and shut off at 175 PSI, you can set the fire pump to start at 155 PSI. That's five PSI above the churn pressure. When the pump starts, the underground pressure is still above the churn pressure, so the check valve on the discharge side of the pump doesn't even open. As the pressure in the underground drops, the check valve opens at 150 PSI. There is no pressure surge because the underground pressure is the same as the pump's churn pressure. And, of course, you already know to arrange the fire pump for a manual shutdown."

Chapter 9 – Special Extinguishing Systems

#33 – Terminology: Suppression vs. Extinguishing

"Hey, Boss," asked the Brash Young Fire Protection Engineer (BYFPE) of his supervisor, the Wise Old Fire Protection Engineer (WOFPE), "Why do you insist that I use the term special fire extinguishing system in my reports instead of special fire suppression system? You made that change about five different places in the last report I submitted for your review."

"Simple," explained the WOFPE. "By definition, they aren't fire suppression systems. They are fire extinguishing systems. A fire suppression system, like an automatic sprinkler system, is designed to suppress or control a fire until fully extinguished, often by the fire department. Look at NFPA standards like NFPA 17 for dry chemical systems, NFPA 2001 for clean agent systems, or NFPA 12 for carbon dioxide systems. The titles of those documents don't say fire suppression system. They say fire extinguishing system."

"OK," said the BYFPE, "But it seems to me that the terms are interchangeable."

"A lot of people think like that," said the WOFPE, "But look at it this way. Suppose you develop a new type of protection system for deep fat fryers. You take it to Underwriters Laboratories or FM Global to have it tested and listed so you can sell it commercially. Do you think you will get a listing for the system if it doesn't EXTINGUISH the test fires they use?

"Probably not," said the BYFPE.

"Definitely not," clarified the WOFPE. "To obtain the listing, you must extinguish, not suppress, the fire. Do you see why we don't call it a special suppression system? The listing laboratories don't give you points for suppressing the fire for two minutes, five minutes, or even thirty minutes. If you don't put the fire out, meaning extinguish it, you don't pass the test and don't get a listing or approval for your product."

#34 – Can I keep my Halon System?

"Hey Boss," hailed the Brash Young Fire Protection Engineer (BYFPE) to the Wise Old Fire Protection Engineer (WOFPE) as they both entered the training lab. "I need your advice on what to do about an existing Halon 1301 system at the Poly Tech Chemical Company headquarters building."

"Tell me about it," said the WOFPE as he sat down at a table in one of the classrooms.

"Well, it's a big system that protects the corporate IT center. There are more than a thousand pounds of Halon 1301 in each of the main and reserve banks of storage tanks," explained the BYFPE. He continued, "From the record drawings, it was installed in the late 1970s but has been upgraded a couple of times since then with a new air-sampling smoke detection system and a new control unit."

"Sounds as though they've taken good care of it," said the WOFPE. "What's your immediate concern?"

"The corporate risk manager asked if they needed to get rid of it and replace it with something else," said the BYFPE. "Looking at the system, it seems to meet their needs, but with the ban on Halon 1301 that took effect before I was born, I wasn't sure if they needed to replace the Halon 1301 with a new clean agent or not."

"There's no requirement that they replace the Halon 1301," explained the WOFPE. "Halon 1301 production stopped in 1994, but it's still available. Some companies specialize in collecting and recycling Halon 1301 from systems removed from service. They stockpile it for critical applications. A system discharge would be costly, but they could obtain Halon 1301 to refill the system. But keep in mind," continued the WOFPE. "Replacement of the Halon isn't the biggest concern."

"Oh, what's the biggest concern?" asked the BYFPE.

The WOFPE explained, "You said the initial system installation dates to the late 1970s. That's more than forty years ago. First, what condition is the space in right now? Over the past forty years, I would guess they've installed and removed at least ten or twelve generations of cabling and communication lines as they upgraded and changed their IT infrastructure. That room probably looks like Swiss cheese, and it may not hold the Halon if they did have a fire and system discharge."

"Second," continued the WOFPE, "They could still get Halon 1301 if needed, but what about replacement parts for the system hardware? Where would they go if they needed to replace a discharge head that hasn't been manufactured for twenty years? Trying to get parts is the real problem. If something breaks, they could be without protection for a long time."

"I hadn't thought of that," said the BYFPE. "I think I'll recommend a door fan test on the IT space just to see how tight it is. I'll also recommend they start budgeting for a new clean agent system using the recently installed air-sampling smoke detection. And I'll check current pricing to see what they might get from a recycler for the two thousand pounds of Halon 1301 they have. That would offset some new system costs when the time comes."

"Now you're thinking," said the WOFPE. "Don't just dump a problem on them. Provide them a solution as well."

#35 – Time Delays and Abort Switches

"Hey Boss," beckoned the Brash Young Fire Protection Engineer (BYFPE) to the Wise Old Fire Protection Engineer (WOFPE) as the teleconference ended. "Can I bounce a couple of things off of you?"

The BYFPE plopped into a comfortable chair across from the desk and began, "Remember that discussion we had about the Halon 1301 system at the Poly Tech Chemical Company headquarters building a couple of weeks ago? You know, the one installed in the 1970s that's still in service?"

"Yes," answered the WOFPE. "You were recommending they do a door fan test to check the room's integrity and that they start planning to replace the system, correct?"

"That's the one," responded the BYFPE. "The corporate risk manager wholeheartedly agreed with the recommendations and engaged us to develop a Scope of Work for the project."

The WOFPE motioned with his hand for the BYFPE to continue.

"I prepared a preliminary Scope of Work and reviewed it with the Risk Manager and the Chief Facilities Engineer," explained the BYFPE. "They were completely on board with what I proposed, but then they gave it to the IT manager for his review."

"Uh oh, I think I can see what's coming," sighed the WOFPE.

"You've got it," explained the BYFPE. "The IT manager hasn't got a clue about how the air-sampling smoke detection system works, thinks the extinguishing agent is poison gas, wants all kinds of time delays and abort switches, and wants to take the sprinklers out of the space. He says he doesn't need them washing his equipment away if there is a fire."

"Well, don't worry about the sprinklers," said the WOFPE. "I know the Chief Facilities Engineer, and she'll explain that the sprinklers stay for protection of the building. The other stuff you have to address through some education." He continued, "A time delay for evacuating the space before discharging the extinguishing agent is acceptable. Have them conduct a couple of fire drills in the space to see how long it takes for everyone to evacuate. That becomes the time delay before discharge."

"An abort station," explained the WOFPE, "is simply a means for human disruption of the precisely engineered system operating sequence. Think about it. What sense does it make to provide a smoke detection system that can be one hundred times more sensitive than a standard spot-type smoke detector, a system that can detect a fire before there is any visible sign, and then give an individual with no fire protection training the ability to override it?"

"I know," protested the BYFPE curtly and with more frustration than he intended, "but this IT guy has peculiar ideas about how the system needs to operate. In addition to a sixty-second time delay, he wants an ABORT switch to push and stop the discharge of the fire extinguishing system. He said every place he's ever worked had an abort switch, and he demanded that this new system have one. But, unfortunately, I couldn't find anything in the codes about abort switches."

"First," reassured the WOFPE, "take a deep breath, and don't let this guy flummox you. You wouldn't dictate to him how to arrange the IT operations, and he shouldn't be dictating the design of the fire protection systems. You're correct that the codes don't address abort switches because we don't want them. We must often provide them to satisfy people like this, but in general, we don't want to give them a means of impairing system operation."

The WOFPE continued, "When we do provide an abort switch, we want it installed inside the hazard area."

"Inside the IT space?" questioned the BYFPE.

"Yes, inside the hazard," answered the WOFPE. "We also want it to be a 'deadman' type switch that takes constant pressure to keep it engaged. And operation of the ABORT switch shouldn't recycle any time delays. That way, if there is a fire and someone wants to press and hold the abort switch, they must stay in the room with the fire. Eventually, they will decide to leave, and when they release the abort switch, the system discharges and operates as designed."

Chapter 10 – Management Fire Control Programs

#36 – Written Management Fire Safety Programs

"Hey Boss," hailed the Brash Young Fire Protection Engineer (BYFPE) to the Wise Old Fire Protection Engineer (WOFPE). "Can we talk about your note on the report you returned to me?

"And which note would that be? I seem to recall several recently," said the WOFPE with a sly smile.

"I'm talking about the one where you said that all management fire safety programs must be written," explained the BYFPE. "You added that note to my description of how good the programs are at the International Consolidated Paper plant. I explained in my report how good everything looked there, but you said that all their management fire safety programs needed to be codified in writing. So, if they are handling everything, what makes the difference whether they have a written program?"

"Sean Patrick MacGinnis," replied the WOFPE. "Sean Patrick MacGinnis is why all their management fire safety and control programs must be available to all managers and employees in written form."

With a furrowed brow, the BYFPE inquired, "Who the heck is Sean Patrick MacGinnis, and what does he have to do with the management fire safety programs at ICP?"

"When I first started in the field, my training mentor took me to the Amalgamated Machine Tool Company in Mount Haven," explained the WOFPE. "It was the model for a well-cared-for, well-maintained facility. It was constructed in the late 1920s and early 1930s and had woodblock floors as clean and shiny as most surgical suites. The sprinklers, installed at the time of original construction, looked brand new. Every fire door functioned. Every fire extinguisher was meticulously inspected and maintained. They had a 50-year-old electric fire pump that looked like it had just come from the manufacturer's showroom. The facility epitomized "a place for everything and everything in its place." And it was all because of the plant engineer, Sean Patrick MacGinnis."

The WOFPE continued, "Sean was imposing, both physically and intellectually. A big Irishman, he stood about six foot five and probably weighed 295 pounds, but he was built like a linebacker. And he was sharp as a tack. His mere presence was intimidating. Until I got to know him, he scared me to death. He was one of those people whose simple presence made things happen. And heaven help you if you violated his rules for maintaining order and organization in the facility."

Invoking memories of days gone by, the WOFPE added, "I remember surveying the facility with Sean once, and we came across a damaged fire door. I noted it and two or three other items I saw as we toured the plant. Sean made notes as well. The next morning before we sat down with the plant manager for the exit conference, he took me out in the plant to show me that every item on the list was already corrected. At the exit conference, I had nothing to say other than the place looked great."

"That sounds like the perfect situation," interjected the BYFPE.

"It might sound like the perfect situation, but it wasn't," lamented the WOFPE. "A few years later, I was back in the plant to check out a new extinguishing system. The new system was installed incorrectly, and the whole plant seemed to be in chaos."

"You see, Sean retired. And when he did, there was no one to pick up the mantle for implementing and maintaining good fire protection practices because Sean never wrote down any of his rules or procedures. Yet, through all the years, things got done, not because a company fire protection policy or program said so, but through the sheer force of his formidable personality. People knew exactly what he expected, and they did everything they could to please him to prevent incurring his wrath."

"Unfortunately, a company can't inspire good fire safety practices through the efforts of one or two or even a dozen individuals," expounded the WOFPE. "The fire safety programs and practices must be formalized, written, and disseminated to all employees with a clear message from management that they fully endorse the programs and expect compliance because it's in the best interest of the company and everyone working there. Effective management fire safety programs survive personnel changes only if they are written and instilled in every employee from their first day on the job."

"So," contemplated the BYFPE. "I guess this is what you mean when you always say that there must be continuity if the programs are to exist in perpetuity. If they aren't written, there is neither continuity nor perpetuity."

"Exactly," said the WOFPE with a look of satisfaction.

#37 – Fire Department Notification

"Hey Boss, I was looking at the timeline you put together for the fire in the trash compactor at Acme Consolidated Industries," said the Brash Young Fire Protection Engineer (BYFPE) to the Wise Old Fire Protection Engineer (WOFPE). "Did it really take the fire department twenty-seven minutes to respond to that fire?"

"It most certainly did," noted the WOFPE without looking up from the report he was reviewing.

"What the heck went wrong?" continued the BYFPE. "The Mt. Haven Fire Department has a fully staffed station less than a mile from the plant. So, what was the problem?"

"Their actual response time was less than two minutes once they got the call," explained the WOFPE, "but it took twenty-five minutes for someone to finally call them."

"You're kidding, right?" inquired the BYFPE.

"Afraid not," said the WOFPE. "Worse yet, it was a passerby in a car that saw the smoke and called the fire department. When the first engine company arrived, they found the building evacuated with everyone dutifully assembled at their designated muster points. Obviously, they have an excellent evacuation plan and have practiced it by having frequent drills. Unfortunately, in all their planning, no one was specifically designated to call the fire department. Over 150 people with cell phones stood outside, taking pictures of the smoke and posting them to social media, but nobody thought to call the fire department. Everybody assumed somebody else must have already done that."

"So, what are they doing so it doesn't happen again?" questioned the BYFPE.

"My recommendation was to revise their procedures to designate the receptionist to call the fire department anytime there is a fire alarm," answered the WOFPE. "I also recommended that their procedures require every department manager also call the fire department whenever the building fire alarm sounds. So what if the fire department gets five, or six, or seven calls, it's a lot better than getting none until it's too late."

#38 – Protective System Impairments

"Hey, Boss," said the Brash Young Fire Protection Engineer (BYFPE) as he poked his head into the office of the Wise Old Fire Protection Engineer (WOFPE). "Remember the problem we had with Amalgamated Motors having ten or fifteen impairments each day when they would keep shutting down carbon dioxide systems because someone had to go into machinery enclosures for maintenance or adjustment of the equipment?"

"Yes, I remember well," said the WOFPE. "Then they would forget to restore the system to service when they finished checking or adjusting so you could walk in the plant at any time and find five or six hazards without protection."

With a big grin, the BYFPE explained, "Well, the new security supervisor solved the problem. He called in Industrial Mutual, their property insurance carrier, and had them do a training program on how to implement their impairment handling and tracking system."

"That's a start, but they've had that program in the plant for years," said the WOFPE. "What makes you think they'll follow the procedures this time and not just keep pencil whipping the records to make the insurance company happy?"

"I think the maintenance people will take it more seriously this time," offered the BYFPE. "The security supervisor added a little twist to the program that makes a big difference."

Looking very skeptical, the WOFPE probed, "And what little twist did he add that makes so much difference that you think he solved the problem?"

"He revised the procedures to require the maintenance department technicians to fill out a special form each time they want the CO2 protection for a hazard impaired," explained the BYFPE. "The form has all the standard precautions they must take and requires the maintenance technician to detail the machine for which they want the protection shutdown, the date and time of the shutdown, the reason, and the expected duration. Then right above the signature line, he added a statement that says, I understand that my name is being provided to the Industrial Mutual Insurance Company as the person responsible for having the fire protection system shutdown and that in the event of a fire, I might be held personally accountable for damages and other costs associated with the event."

"It's a pretty effective attention-getter," continued the BYFPE. "The safety supervisor said that shutdowns originally scheduled to last all day suddenly last only fifteen minutes. In some cases, the maintenance tech has them shut the system down and then says, 'Don't go any place. Just stay right here. This will only take me five minutes.' Then the maintenance crew goes in, makes their adjustment, comes back out, and has them restore the system to service. No more having five or ten systems shutdown for hours each day."

"He knows the statement is bogus, and there is no way they could hold the maintenance techs personally responsible. But, he said if it reduces the duration of an impairment, he'll take it," concluded the BYFPE.

"We know that management fire safety programs, like an impairment handling program, don't work without buy-in from everyone," said the WOFPE. "I guess making the employees feel some responsibility is one way of accomplishing that."

#39 – Controlling Contractors

"Come in for a second," beckoned the Wise Old Fire Protection Engineer (WOFPE) to the Brash Young Fire Protection Engineer (BYFPE). "Let's discuss the loss investigation you did for the construction site fire at Universal Consolidated Industries last week."

"You classified the cause of the fire as unsupervised welding operations," said the WOFPE, "but there are no recommendations for how to prevent it from occurring again."

The BYFPE sat down and explained, "It was a mechanical contractor that started the fire, not Universal Consolidated personnel. Universal has a great hot work program, so I didn't make any recommendations for changes to it. If the contractor had been using Universal's program, the fire probably wouldn't have occurred, or there would have only been minor damage."

"OK," said the WOFPE. "Then that's your recommendation."

"What do you mean?" queried the BYFPE. "Are you saying I should recommend that the contractor be required to use Universal's hot work permit system? How does that work?"

"No," stated the WOFPE. "Not just use their hot work permit system. They should be required to comply with ALL of Universal's fire safety programs and practices. It should be a standard clause in any contract Universal awards that the contractor must adhere to all fire protection policies and procedures implemented by Universal."

"Really?" asked the BYFPE. "Will that work?"

"It will, and it does," offered the WOFPE. "Give the fire protection coordinator at Poly Tech Chemicals a call. I've seen how they make it work, and they don't have problems like this."

"What's their secret?" inquired the BYFPE.

"They have a 'take no prisoners' attitude toward controlling the actions of the contractors they hire," explained the WOFPE. "A few years ago, I reviewed the construction of a new building with the site fire protection coordinator when he noticed some welding operations underway on the construction site. We walked over, and he asked to see the contractor's hot work permit. They fumbled around pretending to look for one but didn't have one."

"Apparently," continued the WOFPE, "this was the second time he had caught them conducting hot work operations without a valid hot work permit from the Poly Tech Chemical security department. The fire protection coordinator ordered all work stopped, and then we headed to the Offices of the Chief Facilities Engineer. The fire protection coordinator explained what had happened and that it was the second time the contractor had violated company policies. He also pointed out that it violated their contract, which included a clause requiring the contractor to abide by all Poly Tech Chemical fire safety procedures and regulations."

"So, what happened?" queried the BYFPE.

"This all started about nine o'clock in the morning," explained the WOFPE. "By two o'clock that afternoon, the contractor had loaded up all their tools and equipment and was escorted out the front gate by security. By the time the gate closed behind them, the word had already gone out to all other Poly Tech Chemical facilities worldwide that they were not to hire this contractor because they would not abide by the rules. And word spread through the construction industry faster than flying welding sparks. Making an example of these guys effectively showed every potential contractor that Poly Tech Chemical means what they say. If you want to do work for Poly Tech Chemical, as part of your contract, you must sign an acknowledgment that you received a copy of their fire safety policies and programs, that you will follow them to the letter, and that you understand any violation is cause for immediate termination of the contract. They rigidly enforce their fire safety programs, which pays off for them in the long run."

Chapter 11 – Pre-Emergency Planning

#40 – Fire Brigade Name Games

"Let's talk about your report section on the fire brigade at Ajax International," said the Wise Old Fire Protection Engineer (WOFPE) as he motioned the Brash Young Fire Protection Engineer (BYFPE) into a chair in his office.

"What's to talk about?" questioned the BYFPE as he plopped into a chair across from the WOFPE. "Ajax International doesn't have a fire brigade. I thought I made that clear in my report. They have an emergency action team."

"Oh, yes. That's right," said the WOFPE. "Let's see here. You said they '...have an emergency action team, not a fire brigade because there are no OSHA regulations for emergency action teams.' Is that correct?"

"Yes," explained the BYFPE. "The plant engineer said last year when they had a lost-time injury OSHA came in to investigate and issued them a citation because employees removed a machinery guard. While there, they snooped around and also cited them for a couple of hazard communication violations. She said they wanted to steer clear of anything regulated by OSHA. She asked us not to mention fire brigades in our report."

"So, now they are calling their fire brigade an emergency action team because they think that gets them around the OSHA fire brigade regulations?" questioned the WOFPE.

"That's what they said," replied the BYFPE with a shrug of his shoulders.

"Does their emergency action team respond to fires in the facility?" asked the WOFPE

"Yes, they do," replied the BYFPE.

And is the emergency action team trained to use portable fire extinguishers?" asked the WOFPE.

"Yes," said the BYFPE.

"I hate to break it to you, but playing word games with the OSHA regulations will likely just cause OSHA to land on them harder next time," explained the WOFPE. "OSHA doesn't judge a fire brigade by the name. If it's an organized team trained to respond to and fight fires, it's a fire brigade, regardless of what they call it. They can call it the Hooterville Volunteer Fife and Drum Corps, but if it's trained to fight fires, it's a fire brigade and subject to the OSHA regulations for fire brigades. Now you need to go make a call to the plant engineer and explain it to her."

#41 – Cooperative Planning with the Fire Department: Who's in Charge?

Sitting with a group of other young engineers in the lunchroom, the Brash Young Fire Protection Engineer (BYFPE) hailed the Wise Old Fire Protection Engineer, "Hey, Boss! Sit with us when you get your lunch. We want to pick your brain for a few minutes."

"OK, what's on your mind?" asked the WOFPE a few minutes later as he sat down at the lunchroom table.

"We were discussing the authority of the fire department when they respond to an industrial facility," explained the BYFPE. "Jed here says that the fire department owns the facility once they arrive on the scene and can do whatever they want. I say that's not the case. Who's correct?"

"Well, you both are to an extent," said the WOFPE. "I'm not an attorney, but depending on the statutes in place in some jurisdictions, when there is a fire, and the public fire department arrives on the scene, they own the building until they deem the situation is safe and return the facility to the control of the owner. During that time, they could, technically, do whatever they please."

"However," continued the WOFPE, "Any competent fire chief is going to work with plant management to work cooperatively during an emergency, so they don't endanger their personnel or the community."

"How do you get the fire department to think like that?" questioned the BYFPE.

"I can give you one good example," said the BYFPE. "When there is a fire at the Amalgamated Motors assembly plant, the fire department shows up at the front gate and asks the plant fire chief what they need. The city fire chief works hand in hand with the plant fire chief."

The WOFPE continued, "That relationship resulted from a fire about ten years ago in the assembly plant paint shop. They had a fire in one of the automated paint booths. Broken paint and solvent lines dumped solvent into the water wash at the bottom of the booth. When the fire department arrived, the exhaust fans in the booth were pulling the flames down into the water wash and up the exhaust stacks, but the deluge sprinklers in the exhaust stacks kept everything cool, and the fire was under control."

"Immediately on their arrival," explained the WOFPE, "the fire department ordered all electrical power to the paint shop shutdown just the way they would for a typical fire in a two-story frame dwelling. Unfortunately, killing power to the exhaust fans allowed the fire to flare up, causing extensive damage to the automated paint spray equipment and the paint booth enclosure. After about twenty minutes, the fire department decided they needed to use foam on the fire since it involved a flammable liquid. Damage to the paint spray equipment and the booth totaled more than $2 million. The reality is that if, after their arrival, they had just stood by and allowed the exhaust system to operate until the burning solvent floating on the wash bath burned itself out, the only real damage would have been the broken solvent lines and some scorch marks on the inside of the spray booth enclosure."

"Wow," said the BYFPE. "Did the plant or the insurance company sue the fire department for malpractice?"

"No," said the WOFPE. Nobody got sued. It's doubtful any lawsuit would be successful, and I don't know that you can fault the fire department for trying to do their job, even if they did make the situation worse in this case."

"So, nothing happened?" questioned the BYFPE.

"Oh, plenty happened," said the WOFPE. "About a month after the incident, Amalgamated Motors scheduled an incident debrief. As a 'courtesy,' they invited the public fire department. It was ostensibly to help the plant fire brigade improve their response to future incidents. The real reason was to educate the public fire department about the paint shop's hazards and multiple protection systems. Over coffee, sandwiches, and cookies, Amalgamated paraded in a series of engineers and experts to go over the paint shop's operations."

"As the engineers described all the high voltages, flammable liquids, and general complexities of the paint shop operations, you could see the firefighters stealing glances at each other. Then, they slowly began to realize the potential dangers they unknowingly walked into during the fire and how lucky they were that no one was hurt or worse. At the end of the 'debrief,' the city fire chief took the Amalgamated fire chief off to the side and said, "Next time we come here for an alarm, just have someone meet us at the gate and tell us what you want us to do." Until that point, they had no idea of the complexity of the hazards in the facility."

"Now, the first due alarm companies from every shift annually tour the facility. New Amalgamated Motors fire brigade members attend classes at the city fire training academy. Each year, they hold a joint drill simulating a fire or explosion in a different part of the facility. Anytime an incident occurs in the plant, the fire department shows up and serves as a backup to the plant fire brigade. That's the way things are supposed to work. No competition, just cooperation."

#42 – The Hazards of Assumptions

"I noted in your survey report for AmalgaTek Metals that their fire brigade only maintains fifty pounds of Class D extinguishing agent on-site," said the Wise Old Fire Protection Engineer (WOFPE) to his younger colleague, the Brash Young Fire Protection Engineer (BYFPE). "With the size of some of the magnesium castings they produce, do you think that's enough?"

"No," replied the BYFPE, "but they said if they need more, they just call the fire department and have them bring their supply of Class D agent and then replace it after the fire."

"Really?" questioned the WOFPE. "Did you talk to the fire department about that plan?"

"Do you think that's necessary?" questioned the BYFPE with a slight note of indignation. "AmalgaTek Metals is good for it. They'll repay the fire department."

"I have no doubt they will repay the fire department," said the WOFPE. "But I doubt that the fire department has fifty pounds of Class D agent in the entire department. Everybody assumes that the fire department has plenty of everything they might need."

"What do you mean?" queried the BYFPE.

"Do you remember that training course we ran here for the chemical industry a few weeks ago?" questioned the WOFPE. The BYFPE nodded yes as the WOFPE continued, "Well, Polynomial Chemical sent their plant fire chief and plant engineer to the course, and they paid for the city fire chief to attend as well. During a discussion on foam systems, we did an example to show how much foam concentrate a large tank fire involving a flammable or combustible liquid might require for extinguishment. For our example, we calculated that for the tank on fire and the enclosing diked area, we would need a minimum of about 1500 gallons of foam concentrate for fire extinguishment."

The WOFPE continued, "We asked how many in the class maintained that much foam concentrate in storage at their facility. The Polynomial Chemical plant engineer spoke up and said we only maintain about 200 gallons of concentrate on-site, but our pre-emergency plan is to call the public fire department to bring their foam stock to fight the fire.

Then we will pay to replenish their supply immediately after the fire. As soon as he said that, the city fire chief turned to him and said, we don't maintain that much foam concentrate in storage. We only carry about ten gallons on each piece of apparatus. We only have about 250 gallons in the whole department. If we need more, we plan to call you guys to see if we can borrow some and then replace it after the fire."

"You're kidding," said the BYFPE.

"I'm dead serious. As closely as they've worked together over the years," explained the WOFPE, "neither the fire department nor Polynomial Chemical ever picked up the phone and asked how much foam concentrate the other had available. Instead, they assumed, and you know how much trouble an assumption can get you into."

"Do you have the direct phone number for the Chief of Operations at the fire department?" asked the BYFPE. "I want to eliminate an assumption currently hanging over my head."

#43 – Coordinating with Emergency Response Agencies

"Hey, Boss," said the Brash Young Fire Protection Engineer (BYFPE) as he settled into a chair across from the Wise Old Fire Protection Engineer (WOFPE). "Do you have a minute to discuss the pre-emergency plans I reviewed for the Amalgamated Chemical Company facility?"

"Sure," said the WOFPE leaning back in his chair. "Fire away."

"Well," started the BYFPE looking at his notes. He then continued with a hint of hesitantly, "it seems that ACC has about every possible type of incident covered in their pre-emergency planning . . ."

"I think I hear a 'but' coming," interjected the WOFPE.

The BYFPE gathered his thoughts and continued, "But, it seems like they depend on the public fire department and outside contractors for about everything. They say the fire department is their primary response capability for fires, hazardous materials, medical, and specialized rescue incidents. And their plan says that for recovery operations, they plan to call in outside contractors. They say they have all the bases covered, but their plans just don't seem up to par with the size of the facility."

"What I hear you saying," said the WOFPE, "is that they have filled in all the blanks on the pre-emergency planning forms but that you don't think they have an effective plan in place."

"Exactly," said the BYFPE. "It seems like they pencil whipped the planning to say they did it, but they've not had any drills or other means to test the plan to see if it works."

"First of all," started the WOFPE, "I know that the fire department in their area is a combination career and volunteer fire department that primarily provides fire and emergency medical response. They are not trained or equipped for hazardous materials response beyond the first responder level or specialized rescue situations such as confined space rescue or high angle rope rescue. Both are potential emergencies at ACC. So, that tells me they never talked with anyone at the fire department. All the fire department will do during a hazmat incident is show up, evacuate the area, keep everyone back, and wait for the regional hazmat team to arrive. Even then, they will simply stabilize the situation so it doesn't endanger the public until it can be cleaned up and mitigated by a professional hazardous materials contractor."

The WOFPE continued, "If they are going to depend on outside resources for emergency response, they need to have agreements with those people before an emergency. It may require a contract in place for an immediate response, so they aren't negotiating it during an emergency. For example, the Petrotech Refinery down the road has a contract with an oil well and oil tank firefighting company to handle any major fires. They pay a yearly retainer for immediate response. Without such agreements in place, facility management will be trying to negotiate a contract literally while the facility burns down."

"That's what I thought," said the BYFPE. "I didn't think you could just pick up the phone and get a complete emergency response team at your front gate in twenty minutes."

"You can't," agreed the WOFPE, "and it's the same for contractors to help with recovery after a fire or other incident. There need to be contracts and agreements in place before the emergency that permit facility personnel to immediately call-in contractors to start making repairs and get the facility back into production. Supervisors must have the authority to make those decisions without the need to run everything up the chain of command. All that needs to be codified in pre-emergency plans before the emergency."

The BYFPE chimed in, "I know what you mean. I investigated a loss a couple of months ago where we ended up with a holiday weekend-long sprinkler system impairment because the maintenance supervisor didn't have the authority to call a sprinkler contractor in on overtime without the permission of the plant manager, who was off fishing for the weekend."

"Exactly," said the WOFPE. "A lot of facilities have pre-emergency plans that are written and placed in a binder for review by the insurance company when they come in but aren't worth a darn because no one has ever checked to see if they are practical for use during an emergency."

Chapter 12 – Fire Hazards

#44 – Sprinklers in IT spaces?

"Hey, Boss," said the Brash Young Fire Protection Engineer (BYFPE) as she gently rapped on the metal office door frame, "Got a minute?"

"Sure, what have you got," said the Wise Old Fire Protection Engineer (WOFPE) as he motioned her in and pointed to a chair across from his desk.

"I completed that fire protection review of the new computer center going into the world headquarters of Amalgamated Acme International. In addition to the clean agent suppression system they are planning, I recommended wet-pipe sprinklers throughout the new IT space. Unfortunately, the corporate IT manager controlling the project called and read me the riot act. I'm surprised you haven't gotten a call from him. He said he wanted to talk to the supervisor of the pinhead that would make such a recommendation."

"Interesting," replied the WOFPE, trying to conceal an amused grin. "Give me the details so I can be ready for his call."

"Well, this is a $30 million project to construct a new cutting-edge information technology center to handle all their sales, accounting, inventory, production scheduling, HR, and every other computer system they operate. Instead of having a half dozen different centers, they are consolidating everything and everyone into this one center with state-of-the-art equipment and software. Three main IT spaces are separated from other areas by two-hour fire-rated construction. Each space will also have an air-sampling smoke detection system to actuate a clean agent fire suppression system discharging into the IT space and below the raised floor," explained the BYFPE.

"Sounds appropriate so far. So, what's his problem?" inquired the WOFPE.

"Well, they commissioned a formal Fire Risk Assessment (FRA) for the new facility as required by NFPA 75. I found the FRA report prepared by our competitor in a document dump on the project they gave me a few months ago. It looks like they did a pretty thorough job on the analysis. One of the final recommendations was to provide wet-pipe sprinklers throughout the new IT spaces. When I reviewed the latest design package, it did not include sprinklers, so I recommended wet-pipe sprinklers following the results of the FRA. The IT guy went nuts. He said, 'I thought we already killed that idea. What is it with you guys? Don't you know you can't mix sprinklers with expensive IT equipment? If one goes off, it will wash the equipment and our operations away,'" explained the BYFPE. "Geeze, Boss, I wasn't trying to stir anything up. I was simply pointing out what their own Fire Risk Assessment concluded."

"Don't worry too much about it. You're on solid footing. Not only does their analysis recommend sprinklers, but organizations that operate some of the largest, most critical, most expensive IT spaces in the world, like NASA, the Department of Defense (DOD), and the Department of Energy (DOE), don't simply recommend sprinklers, they REQUIRE sprinklers in their IT spaces. And in most cases, they use ordinary wet-pipe sprinklers," explained the WOFPE.

"Really?" questioned the BYFPE. I knew they were advocates of sprinklers in their buildings, but I didn't realize they require them in IT spaces."

"They sure do," responded the WOFPE as he reached for a manual on the bookshelf behind his desk and opened it up to a page tagged with a stickie note. "Here's the DOE requirement," he said, pointing to a paragraph in the open notebook.

The BYFPE quickly opened her iPad™ and took a photo of the open page. She put two fingers on the screen and spread them apart, enlarging the portion of the image so she could read –

> *Automatic sprinkler protection for computer equipment areas shall be in accordance with NFPA No. 13, "Sprinkler Systems.". . . The purpose of the sprinkler protection is to limit and control major fire incidents and prevent total destruction of the computer system.*

"Here's the NASA requirement," said the WOFPE pointing to an open page in another book retrieved from his bookcase.

> *10.4.2.1 Automatic sprinkler protection is required for all electronic equipment and record storage areas and shall be installed in accordance with NFPA 13 and Chapter 8 of this document. (Requirement)*

"And here is the Department of Defense requirement," said the WOFPE, laying open another manual on his desk and turning it so the BYFPE could read it.

> *Electronic Equipment Areas must be located in facilities protected by wet-pipe automatic sprinklers. . . . A clean agent fire extinguishing system can be provided in addition to, but not in lieu of, the sprinkler system.*

"Wow, this is great information. I'll file this away for future reference because this discussion comes up every time I have to deal with sprinklers and IT spaces," said the BYFPE. "You have any good info to argue the mistaken idea that sprinklers just 'pop off' by themselves?"

"Well, let's back up a step. You can't win an argument," explained the WOFPE. "You must educate. For example, the IT guy worried about sprinklers just 'popping' off by themselves is operating from a position of ignorance."

The BYFPE grinned at her supervisor labeling the IT manager as ignorant. The WOFPE noted the grin and continued, "I don't mean 'ignorant' in the pejorative sense. I mean it from the standpoint that he simply doesn't know how sprinklers operate. A big part of our job is often educating others about fire protection concepts and how fire protection systems, like sprinklers, operate. Unfortunately, most people's knowledge about automatic sprinklers is limited to the ridiculous misinformation they see on television and in movies. You probably see thousands of sprinklers every week. Have you ever seen one just 'pop off' as they show on television?"

"Never," said the BYFPE, nodding in agreement with her boss.

"Of course you haven't," continued the WOFPE, "And I'll bet you've never been struck by lightning, never been in a plane crash, and never won a major prize in the weekly lotto drawing even though your chances of any of those things occurring are generally greater than a sprinkler operating prematurely."

"Thanks, Boss," the appreciative BYFPE answered. "But what if we 'educate' him and provide those explanations, and he still fights the use of sprinklers in the IT spaces? It is in the best interests of our client to install sprinklers."

The WOFPE leaned back in his chair, laid his glasses on the desk, and laced his hands behind his head. "Easy," he said, "Our actual clients, to whom we report for this project, are the corporate safety and security manager and the corporate facilities engineering manager. They both embrace good fire protection and operate from a position of knowledge and logic, not emotion. They'll ensure the project includes sprinklers after receiving your review letter."

#45 – A Warehouse Fire Success?

"Hey, Boss," hailed the Brash Young Fire Protection Engineer (BYFPE) to the Wise Old Fire Protection Engineer (WOFPE) as she caught up with him in the corridor. "I just read a journal article you wrote on the International Aircraft Corporation warehouse fire a few years ago."

"And, I trust you were suitably impressed," said the WOFPE with a grin.

"Well, it's a good article," explained the BYFPE, "but there's something I can't quite get my head wrapped around."

"Expound," said the WOFPE.

"Your article says the warehouse was designed for all types of aircraft parts. It had automatic sprinklers specifically designed for the storage. It had multiple levels of in-rack sprinklers, and the water supply could provide three times the required fire flow. In addition, your article said the full-time plant fire department had a comprehensive pre-fire plan that they practiced with the public fire department on multiple occasions," explained the BYFPE. "And you said when the fire occurred, twenty-four roof level sprinklers operated, ten in-rack sprinklers operated, and the plant fire department backed up by the public fire department showed up and followed their pre-fire plan to the letter."

"Exactly," said the WOFPE. "Everything went exactly as planned and expected. So, what is it that you can't wrap your head around?"

"If everything went so well," questioned the BYFPE, why was there more than four million dollars in damage? And that is in 1990 dollars."

"Those aircraft parts are expensive," explained the WOFPE. "You package them on pallets and stuff them into a warehouse, and you can have a lot of money tied up in a small load. A single pallet of some products might be a quarter of a million dollars or more. So, with that kind of value, you don't have to damage or destroy too many pallets to add up to four million dollars."

"So, the International Aircraft Corporation was happy and thought the loss of four million dollars in the fire was a success?" asked the BYFPE with a questioning look.

"I don't think they were happy they had the fire, but they were happy that the investments they made in good fire protection paid off. The entire warehouse held more than $200 million of products. They looked at it from the standpoint that they saved $196 million of product, not that they lost four million dollars," explained the WOFPE. "Even when everything goes like clockwork, there will still be damage. The more valuable the commodities in storage, the greater the damage, even during a successfully controlled fire. The fire risk management objective for the warehouse was to limit any fire to a single rack and prevent a warehouse shut down due to a fire. They met those objectives, so yes, it was a success."

#46 – Temporary Storage

The Wise Old Fire Protection Engineer (WOFPE) saw the Brash Young Fire Protection Engineer (BYFPE) pass by his office and called out, "Can you come in for a chat?"

"Sure, Boss. What's up?" said the BYFPE as he slid into the chair opposite the WOFPE.

"I'm reviewing your report from the Acme Supply warehouse you surveyed last week. I noticed several photos show pallets of storage in the aisles between the 25-ft high racks, but I couldn't find any recommendations to eliminate the aisle storage. Could you enlighten me?" requested the WOFPE.

The BYFPE hesitated, then began, "Well, I pointed the storage out to the facility manager. She called the logistics coordinator running the warehouse, who explained they didn't store anything there. They simply stage loads there temporarily until loading onto a truck for shipment. He said they didn't have any other place to stage shipping until the designated truck arrives."

"Oh, so it was all just 'temporary' storage?" asked the WOFPE as he made air quotes around the word temporary. "I guess I shouldn't have any concerns since it's just 'temporary' storage? About how often does this situation occur?"

The BYFPE began to look uneasy. Finally, he quietly replied, "Well, it looked like as soon as they loaded one truck, they started staging the load for the next truck. Now that you mention it, there was storage in the aisles almost all day while I was there."

"And are you aware of some secret test information demonstrating that 'temporary' storage burns differently than 'non-temporary' storage?" questioned the WOFPE. "If you do, maybe we could label all the storage in the warehouse as temporary storage and reduce the overall fire hazard."

The wheels turned in the head of the BYFPE as he formulated a response. "I see your point, but they don't have any other place to stage the orders for shipping. What should I tell them?"

The WOFPE leaned forward and folded his hands on the report he was reviewing. "It does sound like they have a problem," he said. "Be we don't want to compound the problem by not pointing out the hazards of storing stock in the aisles. You know as well as I do that fire testing shows that as little as one pallet load of storage in the aisle can form a bridge of fuel that permits fire to jump the aisle from one rack to another, overwhelm the sprinklers, and burn the building to the ground. It has, and continues to happen far too regularly."

"I see what you mean," responded the BYFPE. "I could explain the problem and suggest they remove a few end racks to make space for a staging area so they don't have to stage storage in the aisles. I suppose that would be the prudent thing to do, but the logistics manager isn't going to like it. I didn't want to look like the bad guy."

"They hired us for our fire protection expertise," explained the WOFPE. "We wouldn't be doing our job if we didn't give them the whole story. You're the bad guy only if you don't fulfill your responsibility to give them your best professional advice. If you don't point out the problems and risks arising from aisle storage, you would be derelict in your responsibility to best serve the client's interests. They may choose to ignore your advice, as clients sometimes do, but at least they would be making an informed decision."

"Got it," said the BYFPE taking in a deep breath. "Let me have that report back, and I'll revise it to include recommendations to eliminate the aisle storage. Then I'll call the facility manager to tell her I'm making the recommendations and why."

The WOFPE extended the report across his desk. As the BYFPE took it, the WOFPE said, "Never worry about being the bad guy. Often people don't like what we have to say because it puts a crimp in what they want to do. But if we give our clients all the facts, whether they like them or not, allowing them to make an informed risk management decision, we've done our job."

#47 – At Least it's partially protected! Not!

"Hey, Boss?" said the Brash Young Fire Protection Engineer (BYFPE) to the wise old fire protection engineer (WOFPE) as he entered the office. "I was on-site today for the start of the investigation into the fire that burned the International Amalgamated Industries warehouse to the ground. Unfortunately, even though the fire started three days ago, it's still burning enough that we won't be able to get close for another few days to start the actual investigation and loss report."

The WOFPE closed the file drawer he was looking through and said, "I could write the loss report for you now if you want me to. I'll just leave blanks so you can fill in when the fire started when the fire department declared it under control, and the actual ignition source. I've already got the other information you need."

"How can you say that? You haven't even been to the scene yet," said the BYFPE.

"Easy," said the WOFPE, "Just look at our last loss prevention survey for the facility. We made recommendations five years ago to either reduce the storage height or increase the sprinkler density. We told them exactly what would happen if they had a fire with the current conditions in the warehouse. Unfortunately, they didn't complete the recommendations, so it's no surprise the place burned to the slab when they had a fire."

"Well, the plant manager is some kind of upset," said the BYFPE. "He says he doesn't understand why he spent all that money on sprinklers if they don't work. He said you told him the sprinklers delivered about eighty percent of the required density, and he doesn't understand why they couldn't hold the fire in check until the fire department got there to put it out."

"And how did you respond?" inquired the WOFPE.

"I tried to explain that sprinklers don't work that way," answered the BYFPE. He continued, "I tried to tell him what you explained to me. Either the sprinklers are designed to control the fire or not. There is no middle ground. There is no 'holding the fire in check.' The fire is getting bigger, or it's getting smaller. If the sprinklers don't match the hazard, the fire grows and spreads beyond the design area. At that point, all bets are off, and the building is lost."

"I'm impressed," beamed the WOFPE. "And how did he respond to your explanation?"

"He said if the sprinklers produced eighty percent of the density we recommended, they should have been able to establish eighty-percent control of the fire," said the BYFPE.

The WOFPE took a deep breath and said, "And as you know, because you've heard me say it a hundred times, there is no such thing as a partially protected warehouse. Either it's fully protected, or it's unprotected. There is no middle ground. If it's not fully protected, it may as well be unprotected because the result is the same, a total loss. Let's hope that IAI takes that lesson to heart since, at a minimum, this lesson will cost them several million dollars."

#48 – Flammable Liquid Cabinets

"Hey, Boss," hailed the Brash Young Fire Protection Engineer (BYFPE) as he rapped on the door jamb of the Wise Old Fire Protection Engineer's (WOFPE) office. "Do you have a minute to talk about flammable liquid cabinets?"

"Sure," said the WOFPE pointing to a chair in his office. "What's giving you heartburn today?"

"I spent this week surveying the research center for Poly-Tech Chemical," explained the BYFPE. "And I have questions about a few things that just don't make sense to me. You've been there several times. Maybe you can explain them."

"I know the facility well, so let's have it," said the WOFPE as he gave the BYFPE the 'come on' sign with his hands.

"First," started the BYFPE, "In many of their labs, they just have wooden cabinets that they use to store safety cans with flammable liquid. Why has no one ever pointed out that they need metal safety cabinets for that type of storage?"

"Because my bright young colleague, they don't need metal flammable liquid cabinets," explained the WOFPE. "What they need and what they have are flammable liquid storage cabinets that comply with NFPA 30."

He continued, "Nothing in the code says they have to be metal. NFPA 30 specifically permits wood flammable liquid cabinets and provides specifications for their construction. Among other things, the specifications in NFPA 30 require a minimum of one-inch exterior grade plywood, rabbeted joints, a raised sill or pan at the bottom to contain spills, and specific means of fastening the panels together. The first time I saw them, probably thirty years ago, I jumped to the same incorrect conclusion as you. I said they weren't code compliant."

"Fortunately, their long-time safety manager pulled me aside and pointed out that not only are wooden cabinets permitted, but NFPA 30 provides the specifications for their construction," explained the WOFPE. "He said at one time they had a full-time carpenter on staff, and when he didn't have much to do, he built those cabinets following the specifications provided in NFPA 30. The plant manager liked them because unlike the bright yellow and red cabinets they purchased, these looked like pieces of fine furniture."

"But how can a wood cabinet protect against a fire?" protested the BYFPE.

"Again, if you look at NFPA 30," counseled the WOFPE, "It explains that the intent is to protect the flammable liquid inside the cabinet for about ten minutes, the time needed for building occupants to escape. It doesn't intend for a flammable liquid cabinet, metal or wood, to prevent involvement of the contents in the fire forever."

"OK," said the BYFPE, "But the other thing I saw was that none of the flammable liquid cabinets were vented. The wooden ones don't have vents, and the metal ones still have the original plug in the vent opening."

"It's a myth that flammable liquid cabinets must be vented to the outside or to anyplace else," explained the WOFPE. "There is no fire protection purpose for venting a flammable liquid cabinet. They work just fine without venting."

Again, protesting, the BYFPE said, "But I've seen them vented, and I've always recommended venting. If they don't have to be vented, why is there a removable plug to vent them?"

The WOFPE smiled, removed his glasses, and laid them on his desk. "Listen carefully to what I said. I said venting is not required, and there is no fire protection reason to vent them. I did not say they cannot be vented."

The WOFPE continued, "There may be a valid reason to vent the cabinet. For example, you might provide venting to remove potentially toxic or harmful vapors from chemicals stored in the cabinet. However, if an owner chooses to vent the cabinet, the venting method cannot compromise the cabinet's performance during a fire. This might require a thermally activated damper on the vent line or some other means of closing the vent when exposed to fire."

"Wow," mumbled the BYFPE dejectedly. "It sure would be nice if you wrote down all the stuff you carry around in your head so I could easily find it when I run into these problems."

The WOFPE picked up his glasses and returned to the report he was reviewing. "All in good time, my friend, all in good time."

#49 – Combustible Dust Testing

"Hey, Boss," hailed the Brash Young Fire Protection Engineer (BYFPE) as she rapped on the door jamb of the Wise Old Fire Protection Engineer's (WOFPE) office. "I've got good news about the testing completed on the dust samples we collected at Amalgamated Food Processing a couple of weeks ago."

"Do come in and tell me about it," said the WOFPE. "After reviewing budget reports all morning, I could use some good news. What does the testing tell us?"

Leafing through the test report she was holding, the BYFPE said, "Well, the dust is combustible, but it's not as bad as we first thought."

"Really?" questioned the WOFPE with a look of surprise. "What do you mean by it's not as bad as we first thought. Things looked pretty bad inside the facility, and I'm unaware of any combustible dust situation that isn't a significant problem."

"Let's see," said the BYFPE, "It looks like all the samples tested were combustible, but the worst-case P_{max} , or pressure developed in an explosion when tested, was only 4.2 bar. That's less than half the P_{max} of wheat flour and corn starch. And the worst-case combustible dust constant, K_{St}, for any of our samples is only 42. If I correctly recall, that's not even twenty-five percent of the K_{St} for corn starch and sugar."

With his eyebrows arched above the rims of the reading glasses perched on his nose, the WOFPE questioned, "So, the good news is that the debris field resulting from a deflagration at Amalgamated Foods will be a half or a quarter of the size of the debris field from a deflagration in a sugar mill? Is that what you mean to say?"

What do you mean?" inquired the BYFPE pensively.

After taking a deep breath and leaning back in his chair, the WOFPE expounded, "The testing just confirms what we already knew. Amalgamated Foods has a combustible dust problem. The test results just provide some idea of the potential magnitude of the problem. Information in that test report will be useful for designing any new protection features, such as explosion vents, that Amalgamated might decide to install. But overall, the information doesn't do much else for us at this point. We knew there was a problem. The testing just provides another means of confirming for Amalgamated that they have a problem that needs immediate attention."

"You said the worst-case P_{max} for any of the samples was only 4.2 bar," continued the WOFPE, using his fingers to place quotation marks around the word 'only.' "Let's see, 4.2 bar is about 60 PSI, give or take. Are the buildings and equipment from which we selected the dust samples designed to withstand an internal pressure of 60 PSI?"

"Not even close," replied the BYFPE sheepishly. "More like one or two PSI."

"So what I meant by my earlier comment about the debris field," explained the WOFPE, "was that the final result of a combustible dust explosion is the same whether it involves sugar, coal, corn starch, or the materials we sampled. It destroys the facility and injures or kills people. The only difference is the size of the debris and how far it scatters from the center of the explosion, or, more likely, explosions plural."

With a pondering look, the BYFPE tentatively asked, "You mean the dust sample testing is just a go/no go test for combustibility?"

"Exactly," explained the WOFPE. "Once we establish that the dust is combustible, we know it can deflagrate. All the other information provided by testing is useful only if we are designing explosion resistant construction, protective features, or explosion suppression systems."

He paused for a moment and then continued, "Actually, there is one other use for the information in the test report. After the explosion, the forensic experts and lawyers will use that information to drag the facility owners and management through the smoldering debris to allege negligence for not addressing the problem before someone was injured or killed."

Chapter 13 – Professional Practice

#50 – Engineering Ethics and Professionalism

"Hey, Boss," said the Brash Young Fire Protection Engineer (BYFPE) as he rapped gently on the door frame of the office of the Wise Old Fire Protection Engineer (WOFPE). "I need to fill you in on a situation I encountered today during my survey of the Megacorp distribution center."

"Did you gather everything needed to start the design of their new fire alarm system?" inquired the WOFPE. "We've got kind of a short fuse on that project."

"No, I didn't," explained the BYFPE. "That's what I wanted to discuss with you. I thought the facility manager might have called you already."

"Not yet," said the WOFPE, "but it sounds like you better fill me in on what happened."

The BYFPE hesitated and then commenced an explanation, "The facility maintenance supervisor and I were surveying the facility and gathering measurements when he asked if he could ask me about the requirements for emergency exits. I said sure, and he proceeded to show me the exit doors around the perimeter of the building. Every exit door had a four-point, key-operated deadbolt lock." He scrolled through some photos on his phone and stopped at one. Then, handing the phone to the WOFPE, he said, "This is what every exit door to the exterior looks like."

"I know our scope of work for this project is limited to the fire alarm system, but there are between 250 and 400 people in this facility depending on the day of the week and the time of day," explained the BYFPE. "With these locks on every door, they essentially have only a single exit from the facility, and that's through security at the main entrance. I told the maintenance supervisor that this type of locking mechanism on an exit door was not permitted and was an immediate danger to the people in the building. He said the facility manager had them installed to prevent employee theft. There were reportedly problems with employees taking stock from the warehouse,

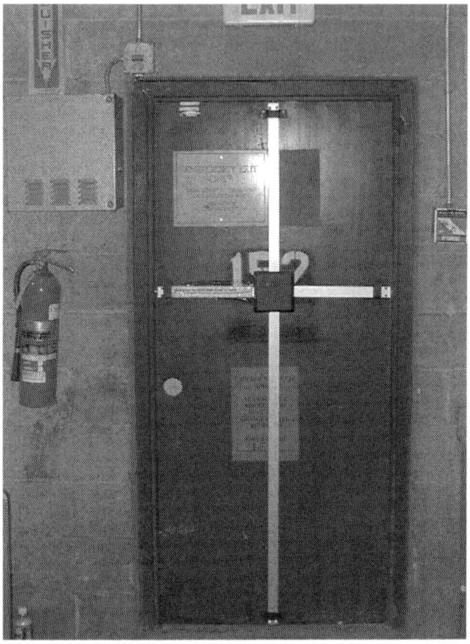

going outside through the exit doors, and throwing it over the fence where friends and family members picked it up."

"I stopped my survey and went to the facility manager to explain the situation to her," continued the BYFPE. "I told her the facility wasn't safe for the people working there and that if the fire marshal's office came in, they could be cited, fined, or even shut down. All I could think of was the fire in that chicken processing plant we studied during training where twenty-five workers died, and over fifty were injured because the plant locked the doors to prevent theft, and the workers couldn't get out during a fire. I said I couldn't continue the survey for the fire alarm system under these conditions. She got agitated and told me that I could just pack up, leave, and don't bother returning. She would find someone else for the fire alarm system. I figured she immediately called you and complained."

The WOFPE laid his glasses on the desk and looked at the BYFPE standing at the door. "Well, I haven't yet heard from her, but don't worry about it. You did nothing wrong. In fact, you did almost everything correctly. You can't just ignore a serious problem because it doesn't fall under your current scope of work. That's not ethical or moral, professionally or personally."

"That's not the complete story," explained the BYFPE with a slight cringe.

"Oh?" said the WOFPE with arched eyebrows, "Maybe you should make like Paul Harvey and tell me the rest of the story?"

"After the facility manager threw me out," explained the BYFPE, "I figured I was already in trouble, but I couldn't just ignore a situation that I thought posed a clear and immediate danger to all the people working in that facility. So, I drove to fire department headquarters, explained the situation to the fire marshal, and showed him my photos. He thanked me, lit out of there like, no pun intended, he was going to a fire and said that before he left, those doors would be open, or the place would be closed. I figured as soon as he showed up at the facility, you'd get a call from the facility manager, and I'd be looking for a new job."

The WOFPE smiled and, with a note of assurance, said, "No need to worry. You've still got a job here, and you've got my admiration for doing it ethically and professionally. No fire protection professional should ever walk away from an obvious hazard just because identifying and correcting it is not in the scope of their work. What is the saying? If you see something, say something. And by the way, I was incorrect when I said that you did almost everything correctly. By contacting the fire marshal to make sure a dangerous condition gets resolved, you did everything correctly. That's exactly what I would do."

#51 – The Fire Protection Engineer as a Fiduciary

After the weekly coordination meeting, the Wise Old Fire Protection Engineer (WOFPE) motioned for the Brash Young Fire Protection Engineer (BYFPE) to remain behind for a minute. As everyone else left the conference room, the BYFPE said, "What's up, Boss?"

"You've been doing an excellent job lately, but I want to ensure we are on the same page concerning project cost containment," explained the WOFPE.

The BYFPE responded, "Well, lately, I've focused on reducing construction costs and trying to make every project as economical as possible for our clients."

"That's very admirable," nodded the WOFPE, "but how about you explain to me your idea of what you mean by economical design and construction."

"OK," started the BYFPE. "On the current project for Polytech Chemical, the specification I wrote leaves it open to the contractor to propose what type of pipe to use for the new sprinkler systems. I've also specified sprinkler densities at the low end of the NFPA 13 curves, so the total required water supply is less, and the overall cost of the water supply is as low as possible," proudly explained the BYFPE.

"It's admirable that you are trying to reduce costs for our clients," explained the WOFPE, "but are you examining long-term costs as well as short-term costs?"

"I'm not sure what you mean," responded the BYFPE.

The WOFPE looked at the BYFPE and asked, "Do you know what a fiduciary is?"

"A fiduciary?" questioned the BYFPE. "Isn't that an advisor or banker or someone who handles money and finances for someone else?"

"Often they are," explained the WOFPE, "but that's not the only type of fiduciary. If you look up the definition of fiduciary, you'll find that it's a person with an ethical or legal relationship of trust with another party or organization. Often that means that someone entrusted with funds, like a financial trust, for safeguarding and managing them while looking out for the best interests of the owner of the trust or funds. But that's not the only type of fiduciary."

"What do you mean?" questioned the BYFPE. "And how does that relate to the practice of fire protection engineering?"

Expounding, the WOFPE explained, "Professionals other than bankers and financial advisors should also act as fiduciaries for their clients. For example, as fire protection engineers, we should act as fiduciaries in our client relationships. They vest their confidence in us and rely on us to provide advice and professional services to them in a manner that looks out for their best interests. For example, specifying the lowest sprinkler density needed to meet minimum code requirements so they can obtain a certificate of occupancy for the building may save the client money up front, but is that looking out for their best interests? What happens when the hazard changes in the future? It generally doesn't decrease, so an increase in hazard would require reinforcement of the sprinklers to meet the challenge of the new hazard. That could be costly. It's the client's decision, but we must ensure we are providing all the information, including current and future costs, needed to make an informed decision based on the site-specific needs and conditions of the facility."

"Think about it," continued the WOFPE. "I take my car to a mechanic that I trust. Over the years, he and his employees have demonstrated they have my best interests at heart. I took a car in once that had a major but repairable problem. The cost to complete the repairs would have been substantial, but they said they could repair it. However, they also explained that while they could fix it, there was a considerable risk that other serious problems resulting from the current issue would later surface, and I could incur substantial additional costs. The repair shop owner carefully explained all the options, risks, and costs and opined that if it were him, he would not repair it based on the age and mileage of the vehicle. He could have just gone ahead with the repairs and charged me several thousand dollars. Instead, He gave me all the options, costs, and his opinion, but he let me make the decision. I decided not to repair the vehicle and his shop lost out on several thousand dollars in business. Still, he gained a customer for life because he acted as a fiduciary and looked out for my best interests. His shop also gained many other customers because I now tell everyone how I trust them and how I'm treated. We want our clients to think of us in the same manner."

The BYFPE thought for a moment and then said, "If I understand what you are saying, I shouldn't automatically recommend the least expensive option for a client if it might cost them more money in the future to upgrade the protection as the hazard changes. So, for example, specifying a density of 0.25 GPM/sq.ft. might cost twenty-five percent more than one designed for 0.15 GPM/sq.ft., but it might serve the client better for a longer time. Or instead of specifying a 1000 GPM pump to meet immediate needs, it might be better to specify a 1500 GPM pump to meet anticipated future needs as well."

"Exactly," said the WOFPE, beaming with satisfaction because of his protégé's newfound understanding of fiduciary responsibility. "We don't make the decisions for our clients, but we should provide them with all the options, the good, the bad, and the ugly of each option, and provide them with a recommended course of action so they can make an informed decision that best meets their long-term needs."

If you found this book informative, please consider leaving a review on Amazon.com and other places where book reviews are collected. Your feedback is valuable to other readers considering purchase of the book, to the author, and to the publisher.

Photo Credits –

Page 25 - National Institute of Standards and Technology
Page 50 - FM Global
Page 89 - Pakphoto

FIREMARK
PRESS

Also by Jeffrey Moore, PE, FSFPE

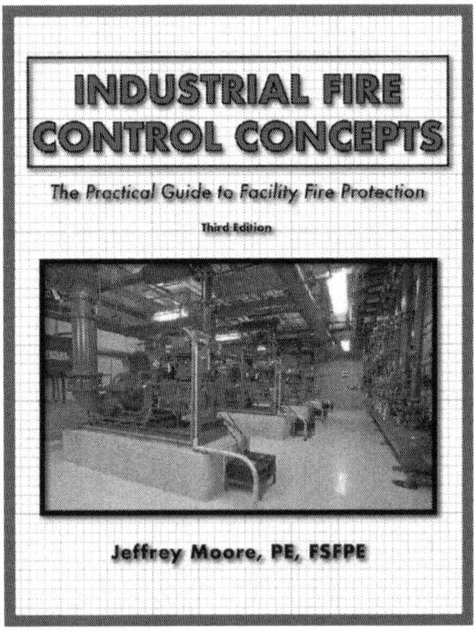

Industrial Fire Control Concepts is a complete facility fire protection course in a book. For more than sixty years, between 1946 and 2007, Industrial Risk Insurers offered an intensive week-long industrial fire protection course in their Hartford (CT) Loss Prevention Training Center and at facilities worldwide. Thousands of facility managers, plant engineers, safety supervisors, insurance loss control engineers, firefighters, government officials, and others benefitted from the course. Although this course concentrated on fundamental fire protection and fire control concepts, participants often felt like they were drinking from a fire hose. They left the course with a thick binder of handouts, articles, notes, references, and other information they could immediately use at their facilities. A common refrain from participants was, "Why isn't all this information available in a book someplace?" In 1988 the first edition of ***Industrial Fire Control Concepts*** attempted to fill this need. This third edition continues the mission of providing a foundation in fire protection concepts allowing the development and implementation of effective site-specific fire protection and fire control strategies for industrial, commercial, and institutional facilities.

This 540 page, full color, heavily illustrated book explains the:

- scope of the fire problem and the lessons of past fire disasters
- basic concepts behind the development and spread of a fire
- holistic systems approach to facility fire safety and fire control
- operation, application, and limitations of various fire protection systems and features
- necessity of a properly arranged fire protection water supply system
- management programs necessary for an effective facility fire control program
- special fire control concerns of information technology (IT) operations, warehousing and storage, flammable and combustible liquids, and combustible dust

The book is available in both print and electronically from Amazon. Download a sample and see how it is suitable for use in nearly any type of fire protection training or education program. For more information or a sample of the text email industrialfirecontrolconcepts@gmail.com

Made in the USA
Columbia, SC
18 June 2025

59537293R00070